# The Michigan Historical Reprint Series

*Reprints from the collection of the University of Michigan University Library*

This volume is produced from digital images created through the University of Michigan University Library's preservation reformatting program. The Library seeks to preserve the intellectual content of items in a manner that facilitates and promotes a variety of uses. The digital reformatting process results in an electronic version of the text that can both be accessed online and used to create new print copies. This book and thousands of others can be found in the digital collections of the University of Michigan Library. The University Library also understands and values the utility of print, and makes reprints available through its Scholarly Publishing Office.

For access to the University of Michigan Library's digital collections, please see http://www.lib.umich.edu.

The Scholarly Publishing Office seeks to disseminate high-quality, cost-effective scholarly content through both print and electronic publishing. Information about the Scholarly Publishing Office can be found at http://spo.umdl.umich.edu.

The Scholarly Publishing Office
the University of Michigan
University Library

# ELEMENTS

OF

# ANALYTICAL GEOMETRY:

EMBRACING

THE EQUATIONS OF THE POINT, THE STRAIGHT LINE,
THE CONIC SECTIONS, AND SURFACES OF
THE FIRST AND SECOND ORDER

BY CHARLES DAVIES.

AUTHOR OF MENTAL AND PRACTICAL ARITHMETIC, FIRST LESSONS IN ALGEBRA,
ELEMENTS OF SURVEYING, ELEMENTS OF DESCRIPTIVE GEOMETRY,
SHADES SHADOWS AND PERSPECTIVE, AND DIFFERENTIAL
AND INTEGRAL CALCULUS.

REVISED EDITION

NEW YORK:
A. S. BARNES & CO., 111 & 113 WILLIAM STREET,
(CORNER OF JOHN STREET.)
SOLD BY BOOKSELLERS, GENERALLY, THROUGHOUT THE UNITED STATES

1867.

TO

# LIEUTENANT-COLONEL S. THAYER,

LATE SUPERINTENDENT

OF THE

## MILITARY ACADEMY.

DEAR SIR,

I take the liberty of inscribing to you the following Treatise on Analytical Geometry, though without flattering myself that the execution of the work will be found answerable to the utility and importance of the subject.

In the organization of the Military Academy under your immediate superintendence, the French methods of instruction, in the exact sciences, were adopted; and near twenty years' experience has suggested few alterations in the original plan.

The introduction of these methods is considered an improvement worthy to form an era in the history of education in this country ; and public opinion has justly appreciated the benefits which you have conferred, at once on the Military Academy, and on the cause of science.

These acknowledgments, prompted alike by a sense of justice and the dictates of private friendship, I have felt it a very grateful duty to make, in prefixing your name to the present work, and in subscribing myself,

With great respect and regard,

Your friend, and obedient servant,

CHARLES DAVIES

Military Academy,

*West Point, July,* 1836.

# PREFACE.

The method, first adopted by Descartes, of representing all the parts of a geometrical figure by a single equation, has wrought an entire change in the mathematical and physical sciences. By means of this happy invention, modes of investigation at once difficult and disconnected, and depending for success in each particular case on the skill and ingenuity of the inquirer, and often on accident, are reduced to a simple and uniform process. The great work of La Place, which from the single law of gravitation deduces the formulas for determining all the circumstances of the solar system, at any period of time, is its legitimate fruit.

In France, much labor and talent have been successfully employed in preparing elementary books on this branch of mathematics, and the schools and colleges are abundantly supplied with those of distinguished merit. In this country, however, it is quite otherwise. No original work has yet appeared, and the translation of one of the most imperfect of the French, and the republication of an English author, can hardly be considered as supplying the seminaries of the United States with suitable text books on so important a branch of science.

For about sixteen years the subject of Analytical Geometry has made a part of the course of mathematics pursued at the Military Academy, and the methods which have been adopted in the present work, are those which have been taught with the greatest success. The admirable treatises of Biot and Bourdon have been freely consulted, and many of the examples in the seventh book have been selected from the former work. The system of Biot has also been somewhat followed. It has been the intention to furnish a useful text-book, and no attempt has been made to depart from clear and satisfactory methods adopted by others, merely for the purpose of seeming to be original.

Military Academy,
*West Point, July,* 1836.

# CONTENTS

## BOOK I.



## BOOK II.

*Ellipse referred to its Conjugate Diameter.*



*Polar Equation of the Ellipse—Measure of the Surface.*



## BOOK V.



*Of the Parabola referred to Oblique Co-ordinates.*

### *Polar Equation—Measure of the Surface.*



## BOOK VI.



### *Of the Hyperbola referred to its Conjugate Diameters.*



### *Of the Hyperbola referred to its Asymptotes.*



### *Polar Equation of the Hyperbola.*

## BOOK VII.



## BOOK VIII.

# ANALYTICAL GEOMETRY.

## BOOK I.

### *Definitions and Determinate Problems.*

1. THERE are three kinds of Geometrical Magnitude: viz. lines, surfaces, and solids. In Geometry, the properties of these magnitudes are established by a course of reasoning in which the magnitudes themselves are constantly presented to the mind. Instead, however, of reasoning directly upon the magnitudes, we may, if we please, represent them by algebraic symbols. Having done this, we may operate on these symbols by the known methods of Algebra, and all the results which are obtained will be as true for the geometrical quantities, as for the algebraic symbols by which they are represented. This method of treating the subject is called *Analytical Geometry.*

Geometry embraces two distinct classes of propositions: viz., problems which relate to particular questions, and theorems which demonstrate general properties.

ANALYTICAL GEOMETRY is also divided into two parts: viz.

1st. The solution, by means of Algebra, of determinate problems: that is, of problems in which the conditions limit the number and determine the values of the required parts: and

2dly The analytical investigation of the general properties of lines, surfaces, and solids.

2. We are first to explain the manner of representing the geometrical magnitudes by the algebraic symbols. For this purpose, it will be necessary to compare each magnitude with

its unit of measure. The unit of measure of any quantity, is a quantity of the same kind, the *value of which is known*, and with which the given quantity is compared. The unit of measure for lines is a right line of a known length, a foot, a yard, a rod, &c. For surfaces, the unit of measure is a known square (Geom. Bk. IV, Prop. IV, sch,); and for solids it is a known cube (Geom. Bk. VII, Prop. XIII, sch).

Let us now suppose that we have a numerical equation of the form

$$x = a + b,$$

in which $x$, $a$, and $b$, are abstract numbers; that is, numbers in which the unit 1 does not express a specific thing. The equation is then called an *abstract* equation.

Let it be now required to find an equation which shall express the same relations between lines as this equation expresses between abstract numbers.

For this purpose, let $l$ designate the unit of measure for lines, and $X$ a line of such a length that it will contain $l$, $x$ times: then,

$$\frac{X}{l} = x,$$

in which $x$ is an abstract number. Let $A$ represent a line which will contain $l$, $a$ times; and $B$ a line that shall contain $l$, $b$ times: then,

$$\frac{X}{l} = \frac{A}{l} + \frac{B}{l},$$

and by multiplying both members by $l$, we obtain

$$X = A + B;$$

that is, the line $X$ is equal to the sum of the lines $A$ and $B$, which is the same relation as subsisted between the abstract numbers of the first equation

This last equation is called a *denominate* equation, because the *kind* of quantity of which its terms are composed, is denominated or named.

Now, since the quotient arising from dividing a quantity by its unit of measure is always an abstract number, it follows that, *any abstract equation may be changed into a denominate equation by substituting for each of the abstract numbers, a geometrical magnitude divided by its unit of measure.*

After these substitutions are made, and the equation cleared of its fractions, all the terms will be homogeneous; that is, each term will contain the same number of literal factors (Alg. Art. 26).

Take, for example, the equation

$$x = ab + c,$$

in which the letters represent abstract numbers. Passing to the denominate equation, we have

$$\frac{X}{l} = \frac{A}{l} \times \frac{B}{l} + \frac{C}{l};$$

or, by multiplying both members of the equation by $l^2$, we obtain,

$$lX = AB + Cl,$$

an equation in which all the terms are homogeneous. As the same reasoning may be applied to every equation, we conclude that, *all the terms of every denominate equation will be homogeneous.*

**3.** A term or factor of the first degree, which represents a right line, is said to be *linear;* and an equation of the first degree is called a *linear equation.*

**4.** When all the terms of an abstract equation are homogeneous, its form will not be altered by changing it into a denominate equation; and we may, therefore, at once treat the equation as though each of its literal factors were linear

Thus, the numerical or abstract equation

$$ax = bc + df,$$

being transformed into a denominate equation, becomes

$$AX = BC + DF;$$

which is of the same form as the given equation.

5. A single factor may always be considered as representing a line; the product of two factors as representing a surface; and the product of three factors as representing a solid. If there are more than three factors in a term, we are to consider only three as linear, and the rest as numerical. Thus, in the equation

$$abx = cdfg + hnm,$$

we may regard either three of the factors $c$, $d$, $f$ or $g$, as linear, but one, at least, must be treated as numerical.

## *Of the construction of Equations.*

6. The construction of an equation consists in finding a geometrical figure in which the parts shall be respectively represented by the literal parts of the equation, and in which the relation between the parts shall be the same as that expressed by the equation.

7. Let it be required, for example, to construct the linear equation

$$x = a + b.$$

Draw an indefinite right line, $AB$. From any point as $A$, lay off a distance $AC$ equal to $a$, and then from $C$, a distance $CD$ equal to $b$, and $AD$ will be the right line equal to $x$

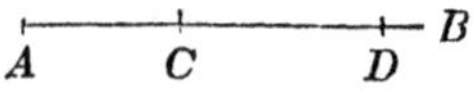

A C D B

**8.** Let it be required to construct the linear equation

$$x = a - b.$$

Draw an indefinite right line $AB$. From any point as $A$, lay off a distance $AD$ equal to $a$, and then from $D$, a distance $DC$ in the direction towards $A$, equal to $b$; $AC$ will then express the difference between $a$ and $b$, and will consequently be equal to $x$.

If $b$ is greater than $a$, $x$ will be essentially negative (Alg. Art. 85); and in the construction, the point $C$ will fall on the left of $A$, leaving for the result, the line $AC$ estimated from the origin $A$ to the left.

We see, in this example, the application of a general principle, which we shall have frequent occasion to verify: viz., *if lines drawn from a given point in one direction are regarded as positive, those drawn from the same point in the contrary direction must be regarded as negative.*

9. Let it be required to construct the equation

$$x = \frac{ab}{c}.$$

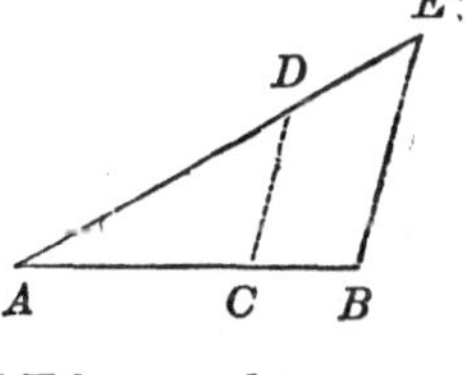


Draw two indefinite right lines $AE$, $AB$, making an angle with each other. From $A$, lay off a distance $AC = c$, also the distance $AB = a$. Then from A, lay off $AD = b$; join $C$ and $D$, and through $B$ draw $BE$ parallel to $CD$; then will $AE$ be equal to $x$.

For, we have by similar triangles,

$$AC : AB :: AD : AE;$$

or

$$c : a :: b : x.$$

Hence, the line $AE$ is represented by $x$.

10 Let it be required to construct the equation

$$x = ab.$$

Since the terms of this equation are not homogeneous, we must pass to the denominate equation, which gives

$$\frac{X}{l} = \frac{A}{l} \times \frac{B}{l},$$

or $$lX = AB.$$

Hence, we see that $X$ is a fourth proportional to $l$, $A$, and $B$

In the last equation, $X$ has the same numerical value as $x$ in the given equation, and since $l$ is the unit of length, the product $lX$ contains the same number of units as $X$. But the units of $x$ are abstract units 1; those of $X$ are units of length; and those of $lX$ are units of surface.

11. Let it be required to construct the value of $x$ in the equation

$$x = \frac{abc}{df}.$$

The equation can be placed under the form

$$x = \frac{ab \times c}{d \times f}.$$

First, find a fourth proportional $g$ to the three quantities $d$, $a$, and $b$, that is, make

$$d : a :: b : g \quad \text{which gives} \quad g = \frac{ab}{d}.$$

We then have, $$x = g \times \frac{c}{f}.$$

From which we see, that $x$ is a fourth proportional to $f$, $g$, and $c$. In the same manner we might construct all equations similar to the above.

12 Let it be required to construct the value of $x$ in an equation of the form

$$x=\frac{abc+dfg}{hm}.$$

it may be put under the form

$$x=\frac{abc}{hm}+\frac{dfg}{hm},$$

and each term constructed separately; the sum of the separate results will be the value of $x$.

13. Let it be required to construct the equation

$$x=\sqrt{ab},$$

or $$x^2=ab,$$

in which it is plain, that $x$ is a mean proportional between $a$ and $b$.

Draw an indefinite right line $AB$, and from any point as $A$ make $AB=a$, and then $BC=b$. On $AC$ as a diameter describe a semicircle, and from $B$ draw $BD$ perpendicular to $AC$; then will $BD$ be the value of $x$. For, (Geom. Bk. IV, Prop. XXIII, Cor),

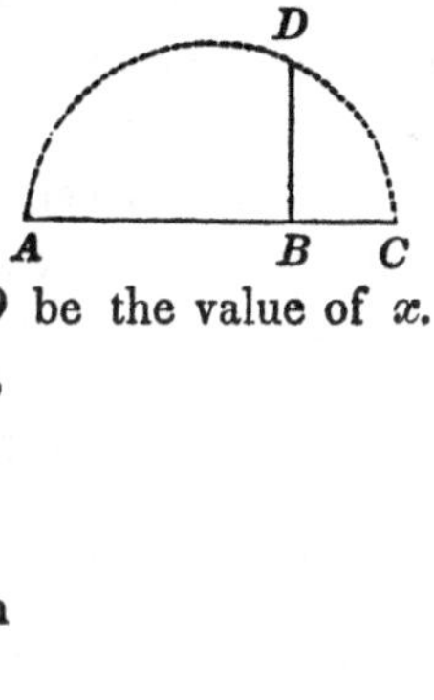

D

A B C

$$\overline{BD}^2=AB\times BC.$$

14. If we have an equation of the form

$$x=\sqrt{a^2+b^2},$$

it is evident that $x$ represents the hypothenuse of a triangle whose two other sides are represented by $a$ and $b$.

15. If we have an equation of the form

$$x=\sqrt{a^2-b^2},$$

it is plain, that $x$ will be the side of a right-angled triangle, of which $a$ is the hypothenuse and $b$ the third side.

16. The methods already explained are sufficient to construct all equations of the first degree, and also all equations of the second degree involving but two terms. We will now explain the manner of constructing the complete equations of the second degree. The first form is (Alg. Art. 144),

$$x^2 + 2ax = b^2.$$

The equation can be put under the form

$$x(x + 2a) = b^2;$$

from which we see, that $b$ is a mean proportional between $x$ and $x + 2a$.

To construct this equation, draw $AB$ and make it equal to $b$. At $B$ erect the perpendicular $BC$ and make it equal to $a$, and join $A$ and $C$. With $C$ as a centre, and $CB$ as a radius, describe a semi-circumference cutting $AC$ in $E$, and $AC$ produced in $D$; then will $AE$ be equal to $x$. For (Geom. Bk. IV, Prop. XXX),

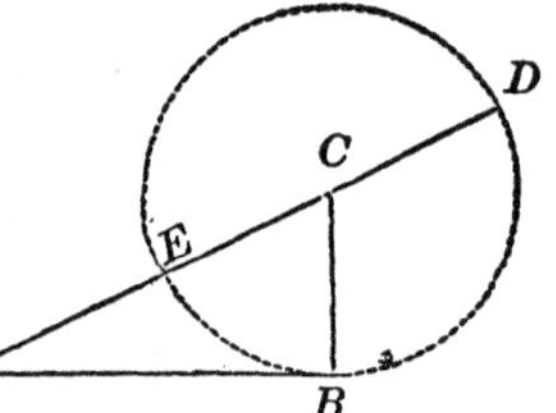


$$AE(AE + 2EC) = \overline{AB}^2 = b^2,$$

$$\text{or} \quad - \quad - \quad x(x + 2a) = b^2.$$

If we solve the equation, we find

$$x = -a + \sqrt{b^2 + a^2}, \quad \text{and} \quad x = -a - \sqrt{b^2 + a^2}.$$

Having described the triangle $ABC$, as before, $AC$ will represent the radical part of the values of $x$.

For the first value of $x$, the radical is positive, and is said off from $A$ towards $C$: then $-a$ is laid off from $C$ to $E$, leaving $AE$ positive, as it should be, since it is estimated from $A$ towards $C$.

For the second value of $x$, we begin at $D$ and lay off $DC$ equal to $-a$; we then lay off the minus radical from $C$ to $A$, giving $-DA$ for the second value of $x$.

Let us now see if this value will satisfy the equation,

$$-x(-x+2a)=b^2,$$

or $$-AD(-AE)=b^2,$$

or $$AD \times AE = \overline{AB}^2.$$

17. The second form of equations of the second degree is

$$x^2 - 2ax = b^2,$$

which gives for $x$ the two values,

$$x = a + \sqrt{b^2 + a^2} \text{ and } x = a - \sqrt{b^2 + a^2}.$$

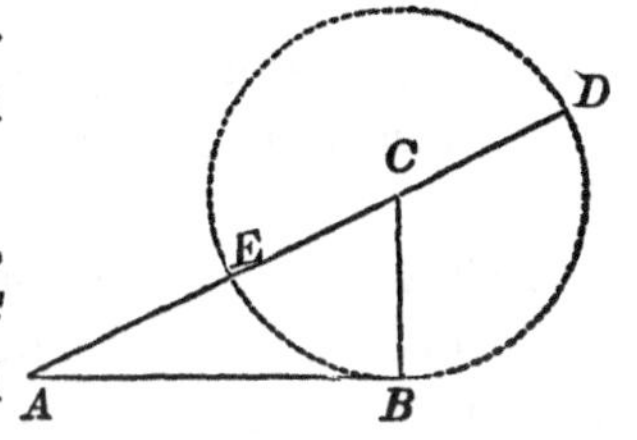


The first value of $x$ is represented by $AD$, estimated from $A$ to $D$.

The second value is $+EC - CA$, the latter being estimated from $C$ to $A$; this leaves $-EA$ estimated from $E$ to $A$.

The positive root in the first construction corresponds to the negative root in the second; and the negative root in the first to the positive root in the second. This is as it should be, since one of the forms changes to the other by substituting $-x$ for $x$.

18. The third form is

$$x^2 + 2ax = -b^2,$$

which gives $x = -a + \sqrt{a^2 - b^2}$ and $x = -a - \sqrt{a^2 - b^2}$.

Draw an indefinite right line $FA$, and from any point as $A$ lay off a distance $AD = -a$, and since $a$ is negative, we lay off its value to the left. At $D$, draw $DC$ perpendicular to $FA$ and make it equal to $b$. With $C$ as a centre, and $CB = a$ as a radius, describe the arc of a circle cutting $FA$ in $B$ and $E$. Now, the value of the radical quantity will be $BD$ or $DE$. The first value of $x$ will be $-AD$ plus $DE$, equal to $-AE$. The second, will be $-AD + (-DB)$ equal to $-AB$: so that both of the roots are negative and estimated in the same direction from $A$ to the left.

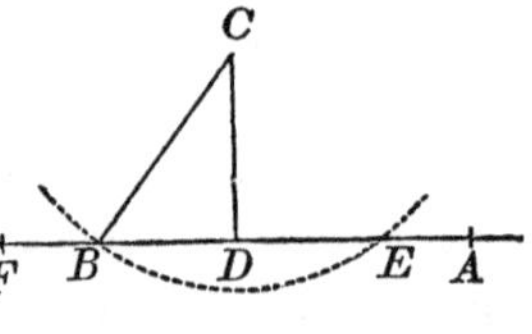


19. The fourth form of equations of the second degree, is

$$x^2 - 2ax = -b^2$$

which gives $x = a + \sqrt{a^2 - b^2}$ and $x = a - \sqrt{a^2 - b^2}$.

Construct the radical part of the value of $x$, as in the last case. Then, since $a$ is positive, we lay off its value $AD$ from $A$ towards the right. To $AD$ we add $DB$, which gives $AB$ for the first value of $x$: and from $AD$ we subtract $DE$, which leaves $AE$ for the second value of $x$. Both values are positive, and are estimated in the same direction from $A$ to the right.

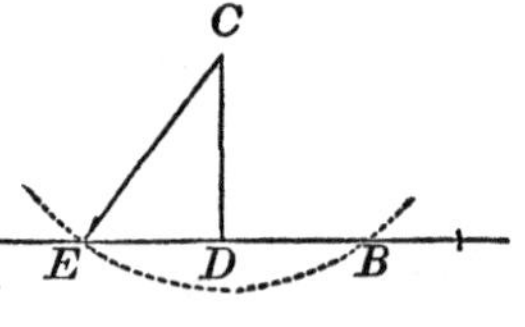


In the two last forms, if $a$ and $b$ are made equal, the two values of $x$ become equal to each other (Alg. Art. 148).

The geometrical construction conforms to this result. For when $a = b$, the arc of the circle described with the centre $C$, will be tangent to $AB$ at $D$, and the two points $E$ and $B$ will unite, and both the roots will become equal to AD.

If $b^2$ be made greater than $a^2$, the value of $x$ in the two last forms will be imaginary (Alg. Art. 147).

The geometrical construction also indicates this result. For, if $b$ exceeds $a$, the circle described with the centre $C$ and radius equal to $a$ will not cut the line $AB$.

Hence, *the imaginary roots of an equation give rise to conditions in the construction which cannot be fulfilled; and this should be so, since the imaginary roots can never appear, unless the conditions of the equation are inconsistent with each other* (Alg. Art. 147).

## *Of Determinate Problems.*

20. No general rule can be given for the solution of geometrical problems. Every new case presents fresh difficulties, which can only be overcome by ingenuity and skill.

In the solution of geometrical problems, by means of Algebra, the following directions may serve as useful guides.

1. Draw a figure which shall represent all the known and required parts of the problem; and then such other lines as may be necessary to establish the relations which exist between them.

2. Represent the known lines by the first letters of the alphabet, $a$, $b$, $c$, $d$, &c.; and the required lines by $x$, $y$, $z$, &c.

3 Consider the geometrical relations which exist between the known and unknown lines of the figure, and express those relations by equations. These equations must be equal in number to the unknown quantities employed.

4. Find, from these equations, the values of the unknown quantities.

5. Construct these values, and unite, if possible, all the lines in a single figure.

PROPOSITION I. PROBLEM.

*Having given the base and altitude of a triangle, it is required to find the side of the inscribed square*

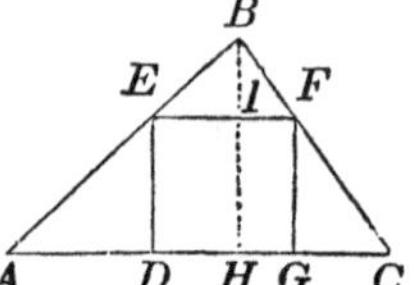


Let $ABC$ be a triangle, in which there are given, the base $AC$, and the altitude $BH$; it is required to find the side of the inscribed square.

Suppose the square $DEFG$ to be inscribed in the triangle, and $BH$ to be drawn perpendicular to the base $AC$.

Designate the base $AC$ by $b$, the perpendicular $BH$ by $h$, and the side of the inscribed square by $x$.

Then, since $EF$ is parallel to the base $AC$, we have, by similar triangles,

$$AC : BH :: EF : BI;$$

that is, $$b : h :: x : h-x,$$

or by placing the product of the means equal to that of the extremes,

$$bh - bx = hx,$$

or $$x = \frac{bh}{b+h}:$$

hence, the numerical value of $x$ is determined, when the values of $b$ and $h$ are known.

But we can also find $x$ by a geometrical construction, since it is a fourth proportional to the three lines $b+h$, $b$, and $h$ (Art. 9). It should, however, be found in such a manner as to connect all the lines with the given and required figures.

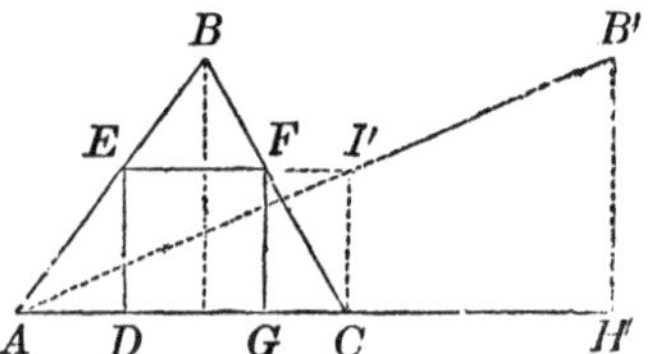


Produce the base $AC$, and on the prolongation lay off $CH'$ equal to the altitude $h$. At $H'$, draw $H'B'$ perpendicular to $AH'$, and make it equal to $h$;

at $C$, draw $CI'$ also perpendicular to the base $AC$. Join $A$ and $B'$, and through $I'$, where the line cuts $CI'$, draw a line parallel to the base: then will $DEFG$ be the required square.

For, we have from the similar triangles $AH'B'$, $ACI'$,

$$AH' : H'B' :: AC : CI' \text{ or } FG;$$

that is, $b+h : h :: b : FG\cdot$

hence, $$FG = \frac{bh}{b+h},$$

and since this relation is the same as that before determined, it follows, that $FG$ is the side of the inscribed square.

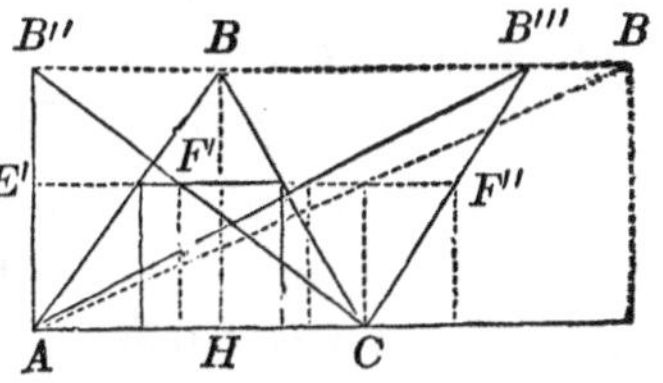

*Scholium* 1. The conditions of the problem fix the value of the base $AC$ of the triangle, and also of the altitude $BH$. They do not, however, determine the position of the vertex $B$. For if $B''B'$ be drawn parallel to $AC$, and at a distance from it equal to $h$, the conditions of the problem would be satisfied by taking the vertex of the triangle at any point of this line, since all the triangles, $AB''C$, $ABC$, &c., would have the same base $AC$, and an altitude equal to $h$. The base and altitude, however, determine the side of the inscribed square, since the triangles $B''AC$, $B''E'F'$, will always be similar: hence, the inscribed square will be equal in all the triangles.

There are, nevertheless, three cases which should be distinguished from each other.

1st. When the angles $A$ and $C$ are both acute, the square will fall entirely within the triangle, as in the triangle $ABC$.

2d. When one of them is a right angle, one angle of the square will coincide with it, and two sides of the square will coincide in direction with the two sides about the right angle, as in the triangle $AB''C$.

3d. When one of the angles is obtuse, the square will fall partly without and partly within the triangle, as in the triangle $AB'''C$: and consequently the problem will be impossible in the strict sense in which it is enunciated.

*Scholium* 2. If in addition to the base and altitude we also know the angles $A$ and $C$, the triangle will then be entirely determined, and the side of the inscribed square may be found by the following construction.

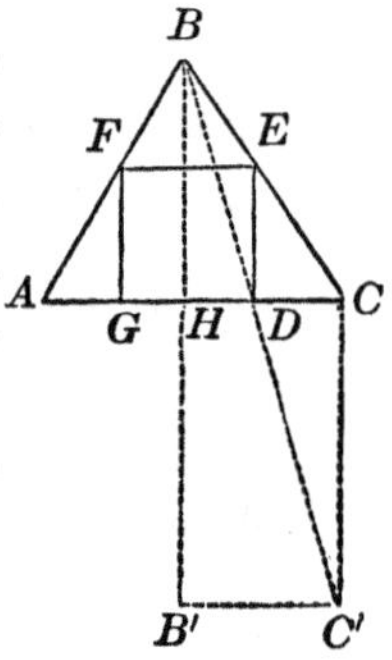

Draw $BH$ perpendicular to the base $AC$, and produce it until $HB'$ is equal to the base $b$. At $B'$ draw $B'C'$ perpendicular to $BB'$, and at $C$ draw $CC'$ perpendicular to $AC$. Through $C'$, the point in which these two lines intersect, draw $BC'$. From $D$, where $BC'$ intersects $AC$, draw $DE$ perpendicular to the base $AC$, and through $E$ draw $EF$, parallel to the base: then will $DEFG$ be the required square.

For, we have by similar triangles

$$BC' : BD :: BB' : BH,$$

and $$BC' : BD :: CC' : ED;$$

hence $$BB' : BH :: CC' : ED;$$

or $$b+h : h :: b : ED,$$

hence $$ED = \frac{bh}{b+h},$$

and therefore $ED$ is the side of the inscribed square.

PROPOSITION II. PROBLEM.

*Having given the base and altitude of a triangle, it is required to inscribe within it a rectangle whose sides shall have to each other a given ratio.*

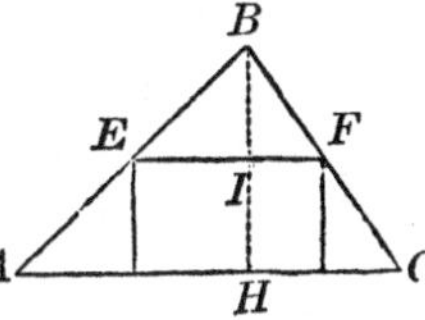


Let $ABC$ be the triangle, having its base $AC=b$, and the altitude $BH=h$.

Let $x$ designate the side of the rectangle which is perpendicular to the base, $y$ the other side, and $n$ the ratio of the sides: that is,

$$\frac{y}{x}=n, \qquad \text{or} \qquad y=nx.$$

Then, from the similar triangles $BAC$, $BEF$, we have,

$$AC : BH :: EF : BI,$$

which becomes $\quad b : h :: y : h-x:$

from which we obtain $\quad bh-bx=hy.$

But we have also found $\quad y=nx.$

Combining these two equations and eliminating $y$, we obtain

$$x=\frac{bh}{b+nh};$$

an expression of the same form as that for the side of the inscribed square in the last problem, excepting that we have $nh$ in the denominator instead of $h$.

But since $n$ is a ratio, it is an abstract number, therefore the expression $nh$ is linear, and implies that the line represented by $h$ is to be taken as many times as there are units in $n$.

We shall give a construction somewhat similar to the last, and which is equally applicable to both problems.

Produce the base $AC$, and on the prolongation lay off $CH'$ equal to $nh$. Through $H'$ draw $H'B'$ parallel to $CB$, and through the vertex $B$ draw $BB'$ parallel to the base $AC$. Join $B'$, their point of intersection, with $A$, and through $F$ draw $FG$ perpendicular, and $FE$ parallel to the base $AC$: then will $DEFG$ be the required rectangle.

For, having drawn $B'P$ perpendicular to $AH'$, we have, by similar triangles,

$$AB' : AF :: AH' : AC,$$

and $$AB' : AF :: B'P : FG;$$

hence, $$AH' : AC :: B'P : FG;$$

that is, $$b + nh : b :: h : FG,$$

or - - - $$FG = \frac{bh}{b+nh}:$$

hence $FG$, or $ED$, is the side $x$ of the inscribed rectangle.

## PROPOSITION III. PROBLEM.

*To draw a common tangent line to two circles in the same plane—their radii and the distance between their centres being known.*

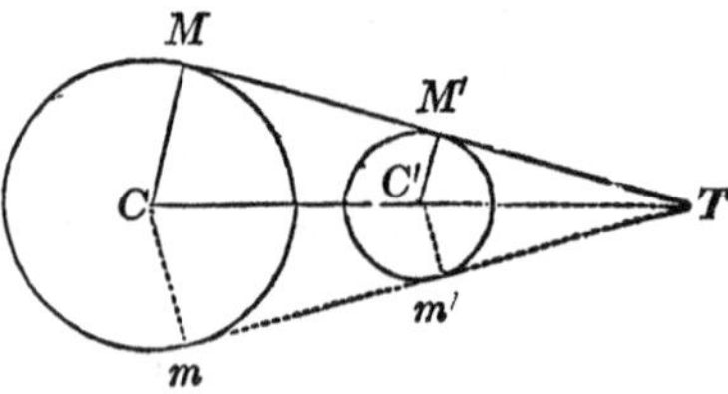


Let $C$ and $C'$ be the centres of the circles, and $CM$, $C'M'$ their radii.

Let the distance betwen their centres be designated by $a$, and their radii by $r$ and $r'$.

Let us suppose the problem solved, and that $MM'T$, or $mm'T$, is a common tangent line to both the circles; and designate the distance $CT$ by $x$.

Now, since the triangles $TMC$ and $TM'C'$ are similar, we have

$$CM : C'M' :: CT : C'T;$$

that is $$r : r' :: x : x-a,$$

or $$rx - ra = r'x;$$

hence, $$x = \frac{ar}{r-r'};$$

from which we see that $x$ is a fourth proportional to the three lines $r-r'$, $a$, and $r$, and this relation will enable us to draw the tangent line.

Through the centres $C$ and $C'$ draw any two parallel radii, as $CN$, $C'N'$. Through $N$ and $N'$ draw the right line $NN'T$, intersecting $CC'$ produced at $T$. Through $T$ draw a tangent to one of the circles (Geom. Bk. III, Prob. XIV), and it will also be tangent to the other.

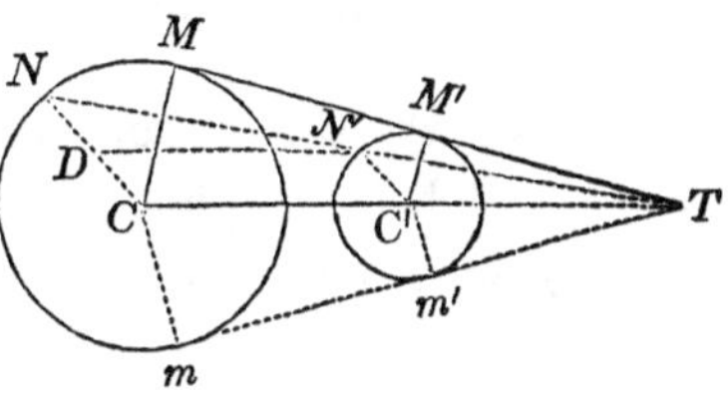


For, through $N'$, draw $N'D$ parallel to $CC'$, and we shall have, by similar triangles,

$$ND : DN' \text{ or } CC :: NC : CT;$$

or $$r - r' : a :: r : CT,$$

hence, $$CT = \frac{ar}{r-r'};$$

the same relation as before established: hence the line $NN'T$ and the tangent line to both the circles, intersect the line $CT$

at the same point; and, therefore, $TM'$, drawn tangent to one of the circles, will also be tangent to the other. Now, since two tangent lines can be drawn through the point $T$ to either of the circles, it follows, that two lines can be drawn tangent to both of them: one on each side of $TC$.

*Scholium* 1. Let us now suppose the larger radius $r$ to remain constant, and the smaller radius $r'$ to increase.

In the equation which expresses the value of $CT$, the numerator $ar$ will remain constant, while the denominator $r-r'$ will continually diminish. When $r'$ becomes equal to $r$, the denominator will become 0, and the value of $CT$ will then become infinite (Alg. Art. 109).

The geometrical construction corresponds with this result: for, when the radii $r$ and $r'$ are equal to each other, the tangent $MM'T$ will be parallel to $CT$, and therefore will not intersect it at any finite distance from $C$.

*Scholium* 2. Let us suppose $r'$ still to increase, in which case it will become greater than $r$. The denominator will then become negative, and since $ar$ is positive, the value of $CT$ will be negative.

The geometrical construction also corresponds with this supposition. For, if $r'$ is greater than $r$, the point $T$ will fall on the left of the centre $C$; and the negative sign merely indicates that $CT$ must be laid off in a direction contrary to that in which its plus value was taken.

*Scholium* 3. There are yet two other tangents which may be drawn to the two circles. These will also intersect the line $CC'$ at the same point, but the point will be between the centres $C$ and $C'$.

Let $C$ and $C'$ be the centres of the circles, $r$ and $r'$ their radii, and $MTM'$ the common tangent line.

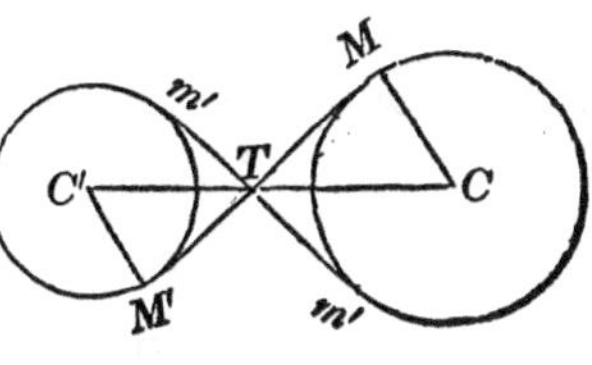


Let the distance $CT$ be again denoted by $x$. Then, since the two right-angled triangles $TMC$, $TM'C'$ are similar, we have,

$$CM : CT :: C'M' : TC',$$

or $$r : x :: r' : a-x;$$

hence $$ar - rx = r'x,$$

from which we find

$$x = \frac{ar}{r+r'},$$

an equation which shows that $x$ is a fourth proportional to $r+r'$, $a$, and $r$.

To make the construction, draw through $C$ and $C'$ the parallel radii $CN$ and $C'N'$, lying on different sides of the line $CC'$. Join $N$ and $N'$. Through $T$, where the line intersects $CC'$, draw two tangents to either of the circles, and they will also be tangent to the other.

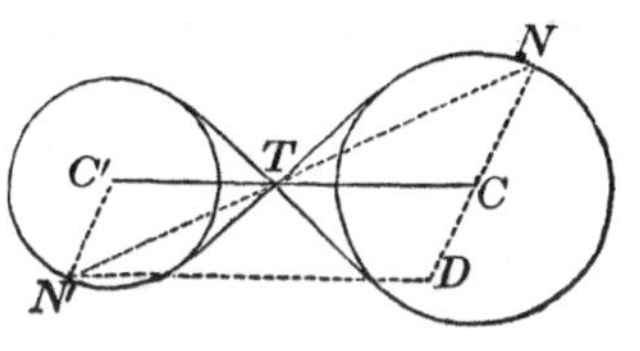


For, through $N'$, draw $N'D$ parallel to $CC'$. Then by similar triangles, we have

$$ND : DN' :: NC : CT;$$

that is, $$r+r' : a :: r : CT,$$

or $$CT = \frac{ar}{r+r'};$$

and as this is the same relation as that before found, it follows that a tangent line drawn through $T$ to one of the circles, will also be tangent to the other.

*Scholium* 4. The problems which have already been proposed, have been resolved by equations of the first degree. We shall now add two that depend on equations of the second degree.

### PROPOSITION IV. PROBLEM.

*To construct a rectangle, knowing the surface, and the difference of two adjacent sides.*

Let $x$ denote the largest side of the rectangle, and $2a$ the difference of the sides: then $x-2a$ will represent the less side.

Let $b$ represent the side of a square equivalent to the surface of the rectangle.

Then, since the surface of a rectangle is equal to the product of its two adjacent sides, we have

$$x(x-2a)=b^2, \text{ or } x^2-2ax=b^2.$$

Resolving the equation, we have

$$x=a+\sqrt{b^2+a^2}, \text{ and } x=a-\sqrt{b^2+a^2}.$$

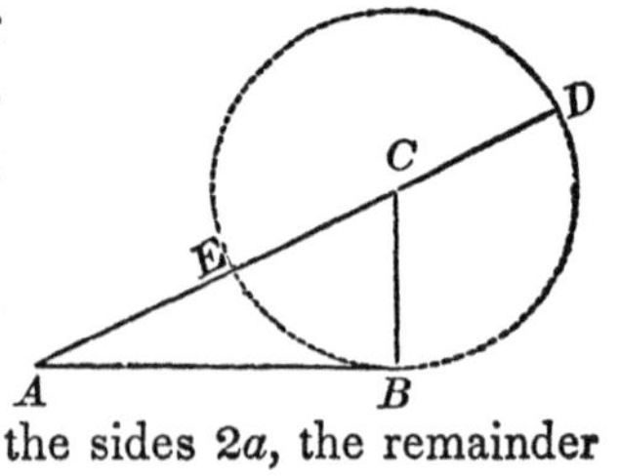

These are the same values for $x$ as found in (Art. 17), and therefore the first may be represented by $AD$ and the second by $-AE$.



Let us first consider the positive value of $x$. If from this value we subtract the difference of the sides $2a$, the remainder will be the less side: that is,

$$a+\sqrt{b^2+a^2}-2a=-a+\sqrt{b^2+a^2}= \text{ less side,}$$

then, $(a+\sqrt{b^2+a^2})\times(-a+\sqrt{b^2+a^2})=b^2$:

hence, the values found for the two sides are verified.

If we take the second value of $x$, we have,

$$a-\sqrt{b^2+a^2}-2a=-a-\sqrt{b^2+a^2}= \text{ the less side.}$$

The term *less* is here to be understood in its algebraic sense, viz: if two quantities are both negative, that is algebraically the less which has the greater numerical value.

The product of the two sides of the rectangle which are found by using the second value of $x$, is

$$(a-\sqrt{b^2+a^2})\times(-a-\sqrt{b^2+a^2})=b^2;$$

hence, these two values for the sides are verified.

The first value of $x$, gave for the sides of the rectangle the two lines $AD$ and $AE$, the product of which is equal to $\overline{AB}^2$: that is to the area of the given rectangle.

The second value of $x$ gives for the sides of the rectangle the two lines $-AE$ and $-AD$, and their product is also equal to $\overline{AB}^2$.

Hence, we see, that either value of $x$ will satisfy the enunciation of the problem understood in its algebraic sense (Alg. Art. 105); for when so understood it is not required that the parts sought should be positive.

*Scholium.* By comparing the two values of $x$, it is seen that the second value taken with a contrary sign expresses the less side of the rectangle. Why ought this to be so?

In every algebraic question, it is proposed to find one or more quantities from the relations which they bear to certain other quantities that are known, and these relations are to be expressed by equations.

Now, if the required quantity be represented by an algebraic symbol, and the conditions of the question be then combined, producing an equation; and if the required quantity be represented by a second symbol, and the conditions again combined producing precisely the same equation, ought not this equation to give the true solution in both cases?

This is precisely what has occurred in the problem. For, we first represented the greater side by $+x$, and found the equation to be

$$x^2 - 2ax = b^2.$$

Had we represented the less side by $-x$, as we were at liberty to do; then

$$-x + 2a = \text{ the greater side,}$$

and $$-x(-x + 2a) = x^2 - 2ax = b^2,$$

the same equation as before.

Now, this equation ought to give not only the greater side of the rectangle which was first represented by $+x$, but also the negative value of the less side, which in the last case is represented by $-x$; and this it does, for we have already shown that the second value of $x$ in the equation is equal to the less side of the rectangle taken with a contrary sign.

The second value of $x$, before its sign is changed, must be treated as the greater side of the rectangle; for, it enjoys the algebraic properties of that side; but, by changing its sign it becomes the less side of the rectangle.

### PROPOSITION V. PROBLEM.

*To divide a straight line into two such parts that the greater part shall be a mean proportional between the whole line and the less part.*

Let $b$ denote the given line, and $x$ the greater part: then will $b-x$ express the less part. Then, by the condition of the question,

$$x^2=b(b-x), \text{ or } x^2+bx=b^2,$$

from which we find

$$x=-\frac{b}{2}+\sqrt{b^2+\frac{b^2}{4}}, \text{ and } x=-\frac{b}{2}-\sqrt{b^2+\frac{b^2}{4}}.$$

To construct these values, draw the indefinite straight line $X'B$, and make $BD$ equal to $b$. At $B$, draw $BC$ perpendicular to $DB$ and make it equal to $\frac{b}{2}$, and draw $DC$. With $C$ as a centre and $CB$ as a radius, describe the circumference of a circle intersecting $DC$ at $E'$ and $DC$ produced in $E$. With $D$ as a centre, and $DE'$ as a radius, describe the arc $E'X$: then will $X$ be the point at which the line $DB$ is to be divided.

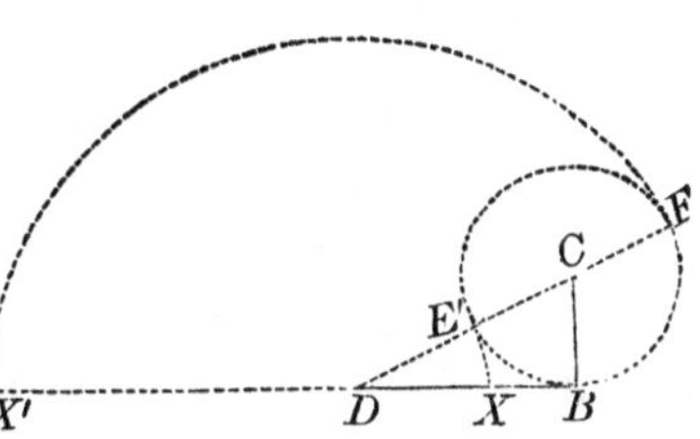


For, the radical part of the values of $x$ is represented by $DC$: hence, the first value of $x$ is represented by $DE'$, and the second, by $-DE$ (Art. 16): therefore, $DE'$ or $DX$ will represent the greatest portion of the given line. To verify it, we have,

$$b-\left(-\frac{b}{2}+\sqrt{b^2+\frac{b^2}{4}}\right)=\frac{3}{2}b-\sqrt{b^2+\frac{b^2}{4}}= \text{ less part.}$$

But we have

$$\left(-\frac{b}{2}+\sqrt{b^2+\frac{b^2}{4}}\right)^2=b\left(\frac{3}{2}b-\sqrt{b^2+\frac{b^2}{4}}\right),$$

since each is equal to

$$\frac{3}{2}b^2-b\sqrt{b^2+\frac{b^2}{4}}.$$

Let us now consider the second value of $x$.

If we square this value of $x$, we have,

$$\left(-\frac{b}{2}-\sqrt{b^2+\frac{b^2}{4}}\right)^2=\frac{3}{2}b^2+b\sqrt{b^2+\frac{b^2}{4}},$$

which is equal to

$$b\left[b-\left(-\frac{b}{2}-\sqrt{b^2+\frac{b^2}{4}}\right)\right]=\frac{3}{2}b^2+b\sqrt{b^2+\frac{b^2}{4}};$$

hence, the second value of $x$ is also a mean proportional between the given line $b$ and the difference between $b$ and the second value of $x$; and therefore this value of $x$ fulfils one of the principal conditions of the question. But it was required to *divide* the line $b$; and since the second value of $x$ is greater than $b$, it cannot fulfil this condition.

But let us enunciate the problem in a general manner: thus, *It is required to find a point on a given line BD, or on BD produced, such, that the distance to D shall be a mean proportional between the distance to B and the given line DB*

If we denote the given line by $b$, and the distance from $D$ to the given point by $x$, we shall have,

$$x^2=b(b-x), \quad \text{or} \quad x^2+bx=b^2;$$

the same equation as before, and which gives the two roots,

$$x=-\frac{b}{2}+\sqrt{b^2+\frac{b^2}{4}}, \quad \text{and} \quad x=-\frac{b}{2}-\sqrt{b^2+\frac{b^2}{4}}.$$

The first value of $x$ gives the distance $DX$, and the second value of $x$, the distance $-DX'$ laid off in a contrary direction: hence, there are two points, $X$ and $X'$ which fulfil all the conditions.

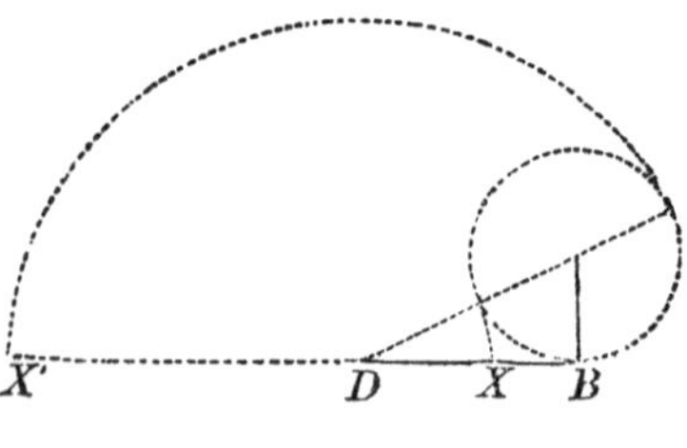

By the first enunciation, the second value of $x$ was excluded from the results, since the conditions required the point to fall between $D$ and $B$. The problem might have been so enunciated as to have excluded the first value of $x$: thus,

*To find a point on the line DB produced, such, that the part produced shall be a mean proportional between the whole line and the given line DB.*

Denoting the given line by $b$ and the part produced by $-x$, we have

$$x^2 = b(-x+b), \quad \text{or} \quad x^2+bx=b^2,$$

the same equation as before. Hence this equation which is the algebraic expression for the three enunciations ought to give the result for each case. Indeed, the first and third enunciations are but particular cases of the second, which is the enunciation of the problem in its most general sense.

We see from this discussion, that a problem may be so restricted in its enunciation as to be solved by one of the roots of an equation of the second degree, and not by the other· and that at the same time a similar problem may be solved by the second root and not by the first. The two problems, however, are so related to each other, that the conditions can be expressed algebraically by the same equation: indeed, they are but particular cases of a more general problem to which each root of the equation is a proper answer.

Having given a sufficient number of examples to indicate the general method of solving geometrical problems by algebra, we shall proceed to the second branch of the subject, viz. the investigation of the properties of lines, surfaces, and solids by means of algebra; and it is this, which, strictly speaking, constitutes the science of analytical geometry

# BOOK II.

## *Of the Point and Straight Line in a Plane—Problems relating to the Straight Line—Transformation of Co-ordinates—Polar Co-ordinates.*

1. In every determinate problem, the conditions of the question limit the number and determine the values of the required parts. Therefore, each algebraic symbol which is employed, represents but a single part of a geometrical figure, and the equations of the problem only express the relations which exist between the given and required parts.

2. When it is proposed to investigate the general properties of geometrical figures by analysis, (and such investigations constitute the science of analytical geometry,) it becomes necessary to assign different values to the same symbol, in order that it may represent, in succession, different parts of the same figure.

Let it be proposed for example, *to find an algebraic expression which shall represent all the points in the circumference of a given circle.*

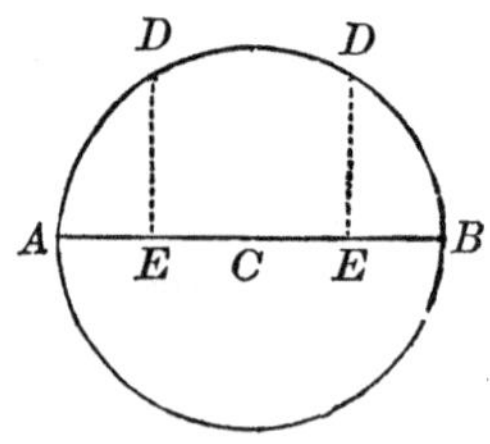


Let $C$ be the centre of the circle, and $CA$ its radius.

Having drawn the diameter $AB$, take any point in the circumference as $D$, and draw $DE$ perpendicular to $AB$. Denote the distance $AB$ by $2r$, the distance $AE$ by $x$, and the perpendicular $DE$ by $y$: then $EB$ will be represented by $2r-x$

Now, $DE$ is a mean proportional between the segments $AE$, $EB$ (Geom. Bk. IV, Prop. XXIII, Cor), that is

$$\overline{DE}^2 = AE \times EB;$$

or by substituting the letters which represent the lines, we have

$$y^2 = x(2r - x), \quad \text{or} \quad y^2 = 2rx - x^2.$$

Since the relation between $DE$ and the parts of the diameter $AE$, $EB$, is the same for any point of the circumference, it follows that this equation may be made to represent each and every point, by assigning to $y$ and $x$ all possible values: that is, all possible values which they can have and at the same time satisfy the equation

$$y^2 = 2rx - x^2.$$

Let us suppose that in this equation we make $x = \frac{1}{2}r$, the equation will then become,

$$y^2 = r^2 - \frac{r^2}{4} = \frac{3}{4}r^2,$$

$$\text{or} \quad - \quad - \quad y = \pm\frac{1}{2}r\sqrt{3}:$$

hence, there are two values of $y$, and these values have contrary signs: which indicates that one is to be laid off above the diameter $AB$, and the other below it.

We might show, in a similar manner, that for every value of $x$, between $A$ and $B$, or between the limits, 0 and $2r$, there will be two corresponding values of $y$ with contrary signs.

If we assign any value to $y$, between the limits 0 and $r$, that is, suppose it equal to a known quantity $b$, we shall have

$$b^2 = 2rx - x^2,$$

which equation will give two values of $x$, and these values can be found since $x$ is the only unknown quantity.

Hence, we see, that, if a particular value be given to $y$, the corresponding values of $x$ can be determined from the equation.

3. The line $DE$, which is represented by $y$ in the equation

$$y^2 = 2rx - x^2,$$

is called the *ordinate* of the point $D$; and $AE$, which is represented by $x$, is called the *abscissa* of the point $D$: and the two taken together, are called the *co-ordinates* of the point $D$. This equation expresses the relation which exists between the co-ordinates of every point of the circumference, and is called the *equation of the circumference*; or simply, *the equation of the circle.* And generally,

*The equation of a line is the equation which expresses the relation between the co-ordinates of every point of the line.*

4. In the equation

$$y^2 = 2rx - x^2,$$

the radius $r$ remains the same for all values that may be attributed to $y$ and $x$. We therefore, call $r$ a *constant* quantity and $y$ and $x$, *variable* quantities.

There are, therefore, two classes of quantities to be considered in analytical geometry;

1st. *The constant quantities, which preserve the same values in the same equation*; and,

2dly *The variable quantities, which may assume all possible values that will satisfy the equation which expresses the relation between them*

These two classes correspond to the known and unknown quantities of determinate problems.

If, in the equation of a line, we attribute a particular value to one of the co-ordinates, the corresponding value of the other will become known.

Thus, in the equation

$$y^2 = 2rx - x^2,$$

if we make $x = r,$

we have $y^2 = 2r^2 - r^2 = r^2,$

or $y = r.$

The reason of this is evident. For, so long as we have one equation and two unknown quantities, there is an infinite number of systems of values which will satisfy it (Alg. Art. 103). But when we attribute a particular value to one of the co-ordinates, we introduce a new condition, and consequently a new equation. The number of equations being then equal to the number of unknown quantities, the remaining co-ordinate ought to be determined in value.

4. By considering the equation

$$y^2 = 2rx - x^2,$$

we see, that $y^2$ will only be equal to $2rx - x^2$, or in other words, the *equation will only be satisfied*, so long as the point $D$ is in the circumference of the circle. For, if it is taken within the circumference, $y$ will be less than $DE$, and we shall have

$$y^2 < 2rx - x^2;$$

and if it be taken without the circumference,

$$y^2 > 2rx - x^2.$$

We will therefore state two propositions, which we shall have frequent occasion to verify.

1st. *If the co-ordinates of any point of a line be substituted for the variables in the equation of the line, that equation will be satisfied.*

2d. *If the co-ordinates of any point, not of the line, be substituted for the variables in the equation of the line, the equation will not be satisfied.*

5. Having shown by a particular example something of the nature of Analytical Geometry, we shall proceed to treat the subject in a more general manner.

In the first seven books, all the points and lines which are considered, are supposed to lie in the same plane. The remaining books will treat of the geometrical magnitudes having any position in space.

6. The terms, *straight line*, and *plane*, will be used, as in Descriptive Geometry, in their most extensive signification. That is, the straight line is supposed to be indefinitely produced, in both directions, and the plane is supposed to be indefinitely extended. When a limited portion of either is to be considered, it will be particularly designated.

7. We shall first explain the manner of expressing, by the algebraic symbols, the position of points and lines on a given plane.

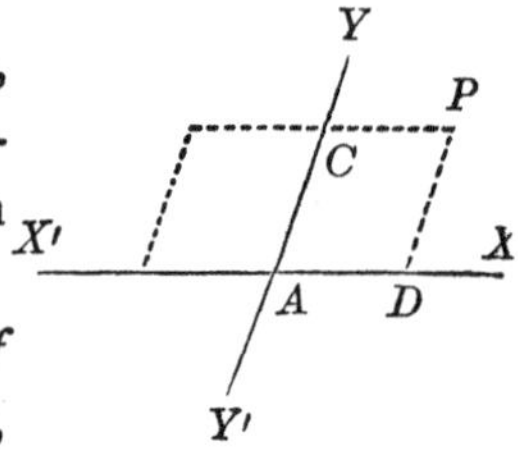

For this purpose, draw, in the plane, any two lines, as $X'AX$, $YAY'$, intersecting at $A$, and making with each other a given angle $YAX$.



The line $X'X$ is called the *axis of abscissas*, or the axis of $X$; and $YY'$, the *axis of ordinates*, or the axis of $Y$. The two taken together are called the *co-ordinate axes*, and the point $A$, where they intersect, is called the *origin of co-ordinates*. The angle $YAX$ is called the *first angle;* $YAX'$, the second angle; $X'AY'$, the third angle; and $Y'AX$, the fourth angle.

8. Let $P$ be any point in the given plane. Through $P$ draw $PD$ parallel to $AY$, and $PC$ parallel to $AX$. Then, $AD$, or $CP$, is called the *abscissa* of the point $P$: $PD$, or $AC$ is called the ordinate of $P$; and the lines $PD$, $PC$, taken together, are called the co-ordinates of the point $P$.

Hence, we see, that the abscissa of any point is its distance from the axis of ordinates, measured on a line parallel to the axis of abscissas; and that the ordinate of any point, is its distance from the axis of abscissas, measured on a line parallel to the axis of ordinates. The co-ordinates may also be measured on the axes themselves. For, $AD$, $AC$, are equal to the co-ordinates of the point $P$.

The co-ordinates of points are designated by the letters corresponding to the co-ordinate axes: that is, the abscissas are designated by the letter $x$, and the ordinates by the letter $y$.

9. If the co-ordinates of a point are given, or known, the position of the point may be found. For, let us suppose that we know the co-ordinates of any point, as $P$. Then, from the origin $A$, lay off on the axis of abscissas a distance $AD$ equal to the known abscissa, and through $D$ draw a parallel to the axis of ordinates. Lay off on the axis of ordinates a distance $AC$ equal to the known ordinate, and through $C$ draw a parallel to the axis of abscissas: the point in which it meets $DP$ will be the position of the point $P$.

When the co-ordinates of a point are known, we have,

$$x = a \quad \text{and} \quad y = b;$$

and these are called, the *equations of the point.*

Hence, *the equations of a point determine its position on the plane of the co-ordinate axes.*

It is evident that, by giving all possible values to $a$ and $b$, the equations of the point $P$ may be made to designate in succession, every point within the angle $YAX$

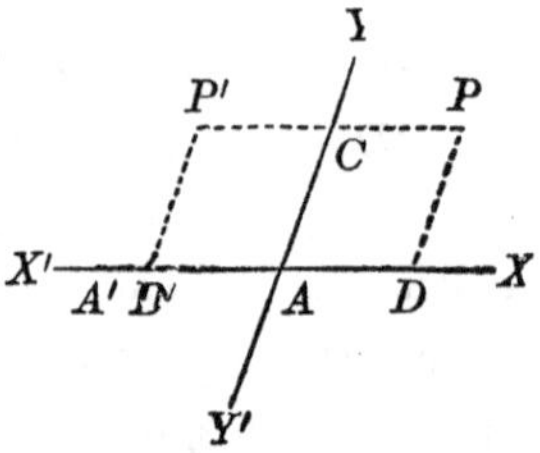

10. Let us now consider a point P′ within the second angle $YAX'$.

The abscissa of this point is $AD'$, and if the points $P$ and $P'$ are equally distant from the axis $Y'Y$, the abscissas $AD$, $AD'$ will be equal to each other. By what notation are these abscissas to be so distinguished from each other that they may both enter into the same general equation? Let us endeavor to explain.

Take any point, as $A'$, on the axis of abscissas, and at a given distance from $A$, and designate that distance by $a$. Let us suppose, for a moment, that $A'$ is a new origin of co-ordinates, and let the abscissas of points referred to this new origin be designated by $x'$.

Now, if we suppose the abscissas which are laid off from $A$ to be designated by $x$, as before, we shall have

$$A'D = A'A + AD; \quad \text{that is,}$$

$$x' = a + x;$$

and also

$$A'D' = A'A - AD';$$

or

$$x' = a - x.$$

If now, we take the first equation,

$$x' = a + x,$$

we see that this equation will express the value of the abscissas $x'$ for every point in the plane of the co-ordinate axes, *provided*, we change the sign of $x$ the moment the point falls on the left of the axis of ordinates $YY'$. Hence, *the abscissas of points may be expressed in a general manner, if those which fall on the right of the origin are regarded as positive, and those which fall on the left, as negative.*

The equation

$$x' = a + x,$$

conforms to this principle with respect to the origin $A'$.

For, if the point $D'$ should fall on the left of $A'$, $x$ would be negative and greater than $a$; the second member of the equation would, therefore, be negative, and consequently $x'$ would be negative, and therefore the abscissas $x'$ are negative when they fall on the left of the origin $A'$.

Since the abscissas of all points in the first and fourth angles fall on the right of the origin, and the abscissas of all points in the second and third angles on the left, it follows, that *the abscissas of all points in the first and fourth angles are to be regarded as positive, and the abscissas of points in the second and third angles, as negative.*

11. Let us now see if the ordinates have similar signs.

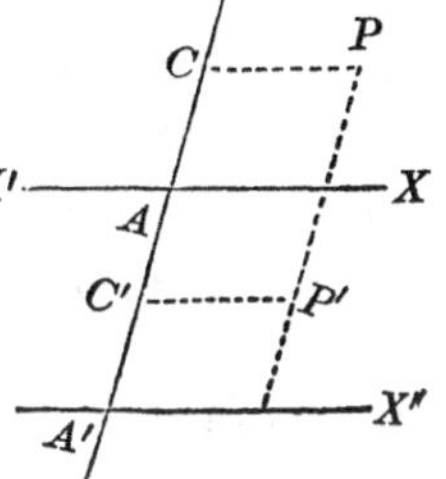


Take any point, as $A'$, on the axis of ordinates, for a new origin of co-ordinates, and denote its distance from $A$ by $b$, and designate the ordinates estimated from the new axis of abscissas $A'X''$, by $y'$.

We shall then have, for the point $P$,

$$A'C = A'A + AC;$$

that is,

$$y' = b + y;$$

and for $P'$, below the axis $XX'$,

$$A'C' = A'A - AC',$$

or

$$y' = b - y.$$

The first equation,

$$y' = b + y,$$

will express the value of the ordinate $y'$ for every point in the plane of the co-ordinate axes, *provided* we change the sign of $y$ the moment the point falls below the axis of abscissas $X'X$. Hence, *the ordinates may be expressed in a general manner, by regarding those which are above the axis of abscissas as positive, and those which fall below it, as negative.*

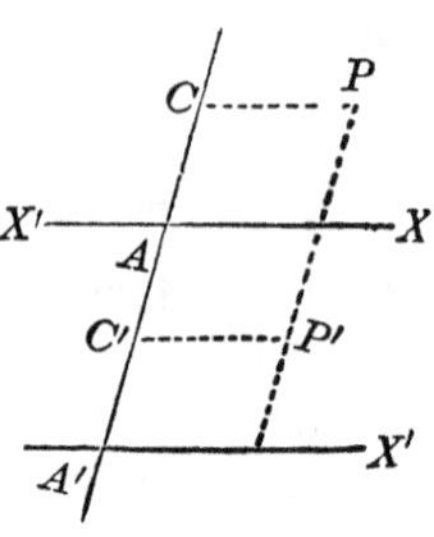


The equation

$$y' = b + y$$

conforms to this principle with respect to the axis $A'X''$.

For, if the point $P'$ should fall below $A'X''$, $y$ would be negative, and greater than $b$; the second member of the equation would, therefore, be negative, and consequently $y'$ would be negative; and therefore, the ordinates $y'$ are negative when they fall below the axis $A'X''$.

Now, since the ordinates of all points in the first and second angles are above the axis of abscissas, and the ordinates of all points in the third and fourth angles below it, it follows, that *the ordinates of all points in the first and second angles are to be regarded as positive; and the ordinates of all points in the third and fourth angles, as negative.*

12. Let us now consider separately, the equations which determine the position of a point in the plane of the co-ordinate axes.

The equation

$$x = a,$$

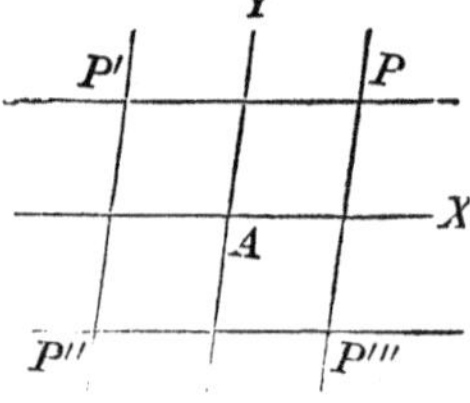


will be satisfied for every point of a straight line drawn parallel to the axis of $Y$, on the right of the origin, and at a distance from it equal to $a$: hence, it will be the equation of that line.

The equation

$$x = -a,$$

is the equation of a straight line, similarly drawn on the left of the origin.

The equation

$$y = b,$$

is the equation of a straight line, drawn above the axis of abscissas, and at a distance from it equal to $b$: and

$$y = -b,$$

is the equation of a line similarly drawn below the axis of $X$

These four straight lines determine, by their intersections the four points, $P$, $P'$, $P''$, $P'''$; one in each of the four angles.

The following are, therefore, the equations of a point in each of the four angles:

| | | |
|---|---|---|
| 1st angle, | $x = +a$, | $y = +b$. |
| 2d angle, | $x = -a$, | $y = +b$. |
| 3d angle, | $x = -a$, | $y = -b$. |
| 4th angle, | $x = +a$, | $y = -b$ |

We see, by inspecting these results, that the signs of the abscissas in the different angles, correspond to the algebraic signs of the cosines in the different quadrants of the circle; and that the signs of the ordinates, correspond to the algebraic signs of the sines (Trig. Art. XII).

13. If in the equation

$$x = a \quad \text{or} \quad x = -a,$$

we make $a = 0$, the line will coincide with the axis of $Y$ Hence, the equations of the axis of $Y$, are

$$x = 0 \quad \text{and} \quad y \text{ indeterminate:}$$

that is, we must be able to assign all possible values to $y$ in order to express every point of the axis of $Y$. If, however, we wish to designate a particular point, we make $y$ equal to its distance from the origin, plus if it is above the axis of $X$, and minus if it is below.

Therefore, the equations of a point on the axis of $Y$, are

$$x = 0 \quad \text{and} \quad y = \pm b.$$

14. If we take the equation

$$y = b, \quad \text{or} \quad y = -b,$$

and make $b = 0$, the line will coincide with the axis of $X$, hence, the equations of the axis of $X$, are

$$y = 0 \quad \text{and } x \text{ indeterminate;}$$

and for a point on the axis of $X$,

$$y = 0 \quad \text{and} \quad x = \pm a.$$

15 The origin of co-ordinates being in both the axes, its equations are

$$x = 0 \quad \text{and} \quad y = 0.$$

PROPOSITION I. PROBLEM.

*To find the equation of a straight line.*

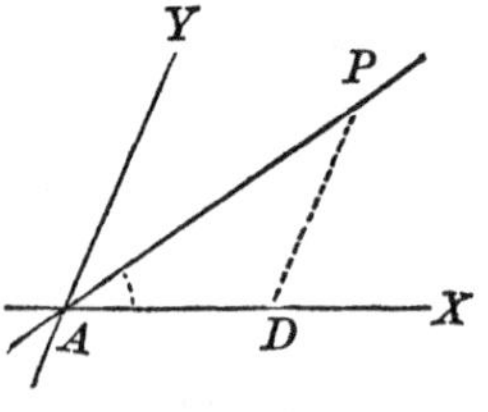


Let $A$ be the origin of co-ordinates, and $AX$, $AY$, the axes. Through $A$ draw any straight line, as $AP$, making with the axis of $X$ an angle equal to $\alpha$. Denote the angle $YAX$ of the co-ordinate axes by $\beta$.

Take any point on the line, as $P$, and draw $PD$ parallel to the axis of $Y$: then $PD$ will be the ordinate, and $AD$ the abscissa of the point $P$.

Since $PD$ is parallel to the axis of ordinates, the angle $APD$ is equal to $PAY$: that is, equal to $\beta-\alpha$.

Now, since the sides of a triangle are to each other as the sines of their opposite angles, we have,

$$PD : AD :: \sin\alpha : \sin(\beta-\alpha).$$

But $PD$ is to $AD$, as any ordinate $y$ of the line $AP$ to the corresponding abscissa $x$: therefore,

$$y : x :: \sin\alpha : \sin(\beta-\alpha),$$

which gives,

$$y=x\frac{\sin\alpha}{\sin(\beta-\alpha)};$$

and this is the equation of the straight line $AP$, since it expresses the relation which exists between the co-ordinates of every point of the line.

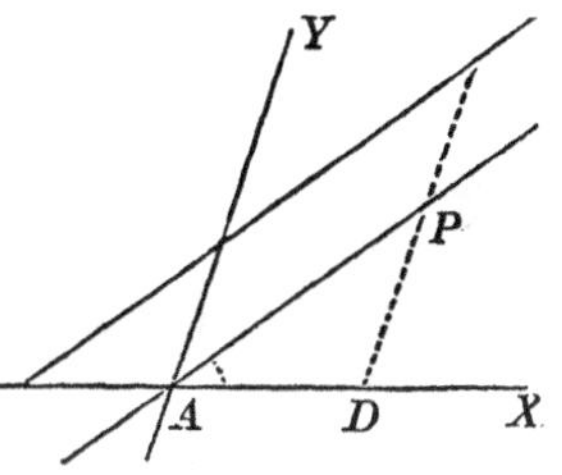


If now, we draw a line parallel to $AP$, cutting the axis of $Y$ at a distance from the origin equal to $b$; it is plain that for the same abscissa $x$, the ordinate $y$ of this new line will exceed the ordinate $y$ of the line through the origin, by the constant quantity $b$: hence, the equation of the last line will be

$$y=x\frac{\sin\alpha}{\sin(\beta-\alpha)}+b.$$

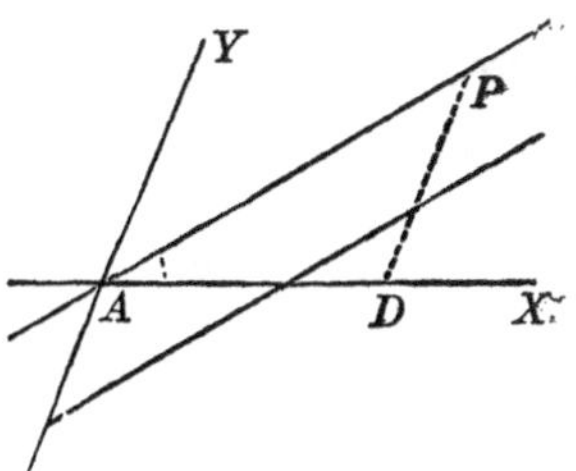


If the parallel cuts the axis $Y$ below the origin of co-ordinates, the value of $y$ in the new line, will be less than the value of $y$ in the line $AP$, by the constant quantity $b$; and in that case, the equation of the parallel becomes

$$y = x\frac{\sin \alpha}{\sin(\beta - \alpha)} - b.$$

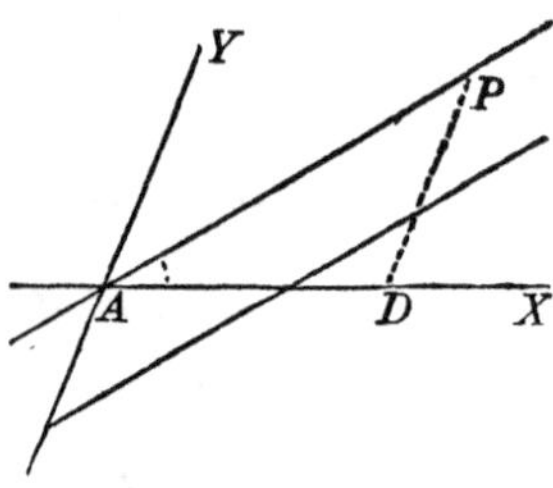


*Scholium* 1. Since the line $PD$ is parallel to the axis of $Y$, the angle $APD$ is equal to the angle $PAY$: hence, *the coefficient of* x *is equal to the sine of the angle which the line makes with the axis of* X *divided by the sine of the angle which it makes with the axis of* Y.

*Scholium* 2. Thus far, we have supposed the co-ordinate axes to make an oblique angle with each other. It is, however, generally most convenient to refer points and lines to co-ordinate axes which are at right angles.

If we suppose $YAX$ to become a right angle,

$$\beta - \alpha = 90^\circ - \alpha,$$

and $\quad \sin(\beta - \alpha) = \cos \alpha \quad$ (Trig. Art. VI).

The equation of the straight line $AP$, passing through the origin of co-ordinates, then becomes

$$y = \frac{\sin \alpha}{\cos \alpha} x,$$

or $\quad y = \text{tang}\, \alpha . x;$

the tangent of $\alpha$ being calculated to the radius of unity.

If we represent the tangent of $\alpha$ by $a$, the equation becomes,

$$y = ax.$$

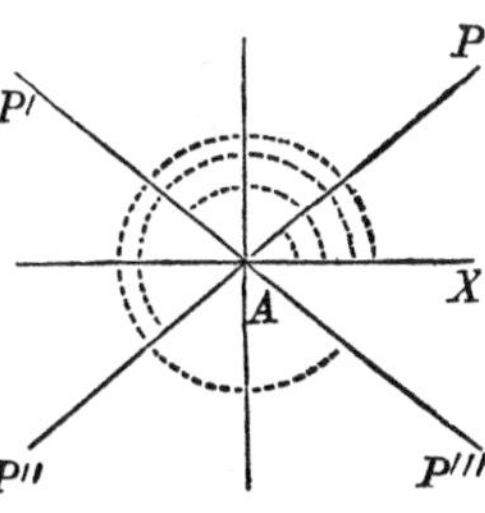


*Scholium* 3. The line $AP$ has been drawn in the first angle. But the equation is equally applicable to a line drawn in either of the other angles, when proper signs are attributed to the tangent $a$, and to the co-ordinates $x$ and $y$. The angle of which $a$ is the tangent is always estimated from the axis $AX$, around to the left, 360°.

If the line, for example, be drawn in the second angle, the angle $XAP'$ will fall in the second quadrant, its tangent, which is $a$, will therefore be negative (Trig. Art. XII). But the abscissas of points in the second angle are also negative: hence, $a$ and $x$ are both negative: their product is therefore positive; hence, $y$ is positive, as it should be since it represents the ordinates of points above the axis of abscissas.

For the line $AP''$, drawn in the third angle, the tangent $a$ will be positive, since the angle falls in the third quadrant (Trig. Art. XII), and since $x$ is negative, the second member will be negative: hence, $y$ will be negative, as it should be.

For the line $AP'''$, drawn in the fourth angle, the tangent $a$ will be negative, since the angle falls in the fourth quadrant, and since $x$ is positive, the second member will be negative, and therefore $y$ will be negative.

Hence, the equation

$$y = ax + b,$$

will represent every straight line which can be drawn on the plane of the co-ordinate axes, if proper values and signs are attributed to $a$ and $b$, and to the co-ordinates $x$ and $y$.

The values of $a$ and $b$ are constant for the same straight line, but take different values when we pass from one line to another. They are often called *arbitrary constants*, because values may be attributed to them at pleasure.

*Scholium* 4. If, in the equation

$$y = ax + b,$$

we make $x = 0$, the value of $y$ will designate the point in which the line intersects the axis of ordinates, for that is the only point of the line whose abscissa is 0. This supposition will give,

$$y = b.$$

If, on the contrary, we make $y = 0$ in the equation of the line, the value of $x$, which is found from the equation, will be the distance from the origin, at which the line intersects the axis of abscissas. This value is,

$$x = -\frac{b}{a}.$$

*Scholium* 5. A line is said to be *given*, or *known*, when the constant quantities which enter into its equation have known values: the position of the line is then determined, and it can be drawn on the plane of the co-ordinate axes.

1. Let us suppose, for example, that in the equation,

$$y = ax + b,$$

the values of $a$ and $b$ are known.

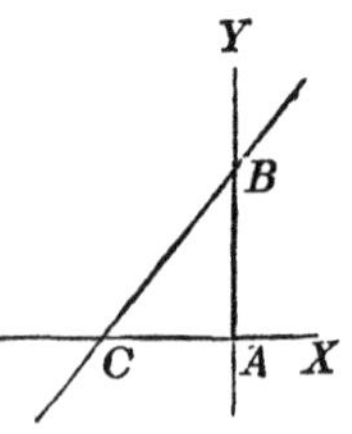


Making $\qquad x = 0,$

we have $\qquad y = b.$

Having drawn the co-ordinate axes $AX$, $AY$, lay off from the origin $A$ a distance $AB$ equal to $b$, and through $B$ draw $BC$, making with the axis of $X$ an angle whose tangent shall be equal to $a$: this will be the straight line whose equation is

$$y = ax + b.$$

The point $C$ might have been found by making

$$y = 0,$$

which would have given,

$$x = -\frac{b}{a} = -AC.$$

If, in the equation of a given straight line,

$$y = ax + b,$$

any value be attributed to one of the variables, the other becomes determinate, and its value may be found from the equation.

If, for example, we make

| | | |
|---|---|---|
| $x = 1,$ | we have | $y = a + b.$ |
| $x = 2,$ | gives | $y = 2a + b.$ |
| $x = 3,$ | gives | $y = 3a + b.$ |
| &c. | &c. | &c. &c. |

Or, we may attribute values to $y$ and find the corresponding values of $x$. If we make

| | | |
|---|---|---|
| $y = 1,$ | we have | $x = \frac{1 - b}{a}.$ |
| $y = 2,$ | gives | $x = \frac{2 - b}{a}.$ |
| $y = 3,$ | gives | $x = \frac{3 - b}{a}.$ |

2. To construct the line whose equation is

$$y = 2x + 5.$$

3. To construct the line whose equation is

$$y = -x - 1.$$

4. To construct the line whose equation is

$$y = -2x + 6.$$

### PROPOSITION II. THEOREM.

*Every equation of the first degree between two variables, is the equation of a straight line.*

The equation

$$Ay + Bx + C = 0,$$

is the most general form of an equation of the first degree between two variables, since there is an absolute term $C$, and since each of the variables, $y$ and $x$, has a co-efficient.

This equation may be written under the form

$$y = -\frac{B}{A}x - \frac{C}{A},$$

which becomes of the form already discussed, if we make,

$$-\frac{B}{A} = a, \quad \text{and} \quad -\frac{C}{A} = b.$$

Having drawn the co-ordinate axes at right angles to each other, if we lay off on the axis of $Y$ a distance equal to $-\frac{C}{A}$, and through the point so determined draw a line which shall make with the axis of $X$ an angle whose tangent is $-\frac{B}{A}$, it will be the straight line whose equation is

$$Ay + Bx + C = 0.$$

We may also put the equation under the form

$$x = -\frac{A}{B}y - \frac{C}{B},$$

in which $-\frac{A}{B}$ is the tangent of the angle which the straight line makes with the axis of $Y$, and $-\frac{C}{B}$ the distance cut off from the axis of $X$, measured from the origin. We may, therefore, state this general principle.

*If, in the equation of a straight line, the coefficient of either variable be made equal to unity, the coefficient of the other variable will be the tangent of the angle which the line makes with the axis of that variable; and the absolute term will be the distance cut off from the axis of that variable whose coefficient is unity.*

## PROPOSITION III. PROBLEM.

*To find the distance between two given points in the plane of the co-ordinate axes.*

A point is said to be given when its co-ordinates are known. Known co-ordinates are usually designated by marking the letters, thus,

$$y', x'; \quad y'', x''; \quad y''', x''';$$

which are read, $y'$ prime, $x'$ prime, $y''$ second, $x''$ second, &c.

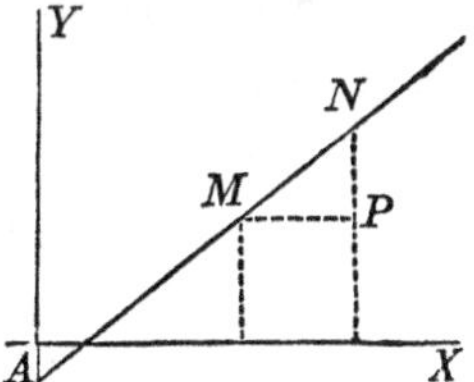


Let $M$ and $N$ be the two given points. Designate the co-ordinates of $M$ by $y'$, $x'$, and the co-ordinates of $N$ by $y''$, $x''$, and the required distance $MN$ by $D$. Then,

$$MP = x'' - x',$$

and $$NP = y'' - y'.$$

But, $$\overline{MN}^2 = \overline{MP}^2 + \overline{PN}^2:$$

hence, $$D^2 = (x'' - x')^2 + (y'' - y')^2,$$

or $$D = \sqrt{(x'' - x')^2 + (y'' - y')^2}; \text{ that is,}$$

*the distance between any two points is equal to the square root of the sum of the squares of the differences of their abscissas and ordinates.*

*Scholium.* If either of the points, as $M$, coincides with the origin, its equations will become

$$x' = 0 \quad \text{and} \quad y' = 0,$$

and we shall have

$$D = \sqrt{x''^2 + y''^2},$$

a result which may be easily verified.

PROPOSITION IV. PROBLEM.

*To find the equation of a straight line which shall pass through a given point.*

Let $M$ be the given point, and designate its co-ordinates by $x'$, $y'$.

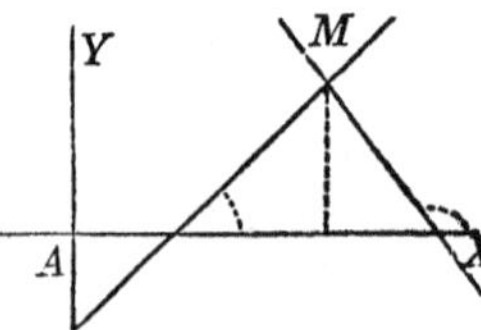


The equation of the line will be of the form

$$y = ax + b,$$

in which $a$ and $b$ are both unknown.

Since the line is to pass through the point $M$, the co-ordinates of this point must satisfy the equation,

$$y = ax + b;$$

hence, we shall have

$$y' = ax' + b;$$

subtracting this from the last equation, we obtain

$$y - y' = a(x - x'),$$

which is the equation of a line passing through the given point, and in which $y'$ and $x'$ are the co-ordinates of the given point, and $y$ and $x$ the general co-ordinates of the line.

*Scholium.* In the equation

$$y - y' = a(x - x'),$$

the tangent $a$ remains undetermined. This is as it should be, since an infinite number of lines may be drawn through the point $M$.

### PROPOSITION V. PROBLEM.

*To find the equation of a straight line which shall pass through two given points.*

Let $M$ and $N$ be the two given points. Designate the co-ordinates of the first by $x', y'$, and the co-ordinates of the second by $x'', y''$.

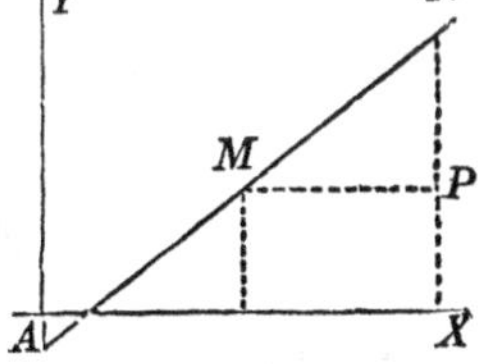


The equation of the required line will be of the form

$$y = ax + b,$$

and it is required to determine the values of $a$ and $b$ in terms of the given co-ordinates $x'$, $y'$, $x''$, $y''$.

Since the required line must pass through the point $M$, the co-ordinates $x'$, $y'$, will satisfy its equation, and we shall have

$$y' = ax' + b;$$

and since it must also pass through the point $N$, we also have

$$y'' = ax'' + b.$$

Subtracting the second equation from the third, we have

$$y'' - y' = a(x'' - x');$$

from which we find

$$a = \frac{y'' - y'}{x'' - x'}.$$

If this value of $a$ be substituted in the second or third equation, $b$ will be the only unknown quantity, the value of which may, therefore, be determined.

It is, however, better to place the required equation under another form.

If we subtract the second equation from the first, we obtain

$$y - y' = a(x - x');$$

Substituting the value of $a$ found above, we have

$$y - y' = \frac{y'' - y'}{x'' - x'}(x - x'),$$

and this line will pass through the second point.

If we subtract the third equation from the first, we shall have

$$y - y'' = a(x - x''),$$

or

$$y - y'' = \frac{y'' - y'}{x'' - x'}(x - x'').$$

*Scholium* 1. The value for $a$, found above is easily verified. For, $y'' - y'$ is equal to $NP$ and $x'' - x'$ is equal to $MP$: hence

$$\frac{NP}{MP} = \frac{y'' - y'}{x'' - x'},$$

and consequently equal to the tangent of the angle $NMP$ to the radius of unity (Trig. Th. II, Cor. 1).

*Scholium* 2. If in the equation for the value of $a$, we suppose $y' = y''$, the value of $a$ will become

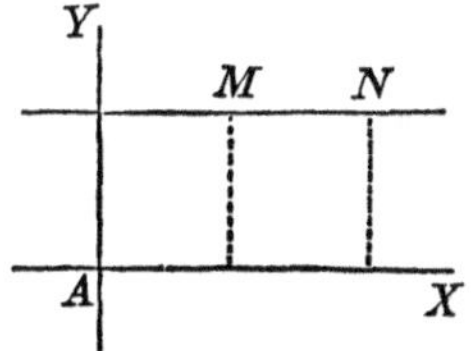


$$\frac{0}{x'' - x'} = 0;$$

and this is as it should be, since under this supposition, the line becomes parallel to the axis of $X$.

*Scholium* 3. If we suppose $x' = x''$, the ordinates $y'$ and $y''$ being unequal, we shall

have $\quad a = \dfrac{y'' - y'}{0};$

therefore, $a$ is infinite (Alg. Art. 109), and hence, the line is perpendicular to the axis of $X$ (Trig. Art. IX).

If we suppose $y' = y''$, and at the same time make $x' = x''$, the two points will coincide, and we shall have

$$a = \frac{0}{0};$$

hence $a$ is indeterminate (Alg. Art. 110), as it should be, since an infinite number of lines can be drawn through a single point.

PROPOSITION VI. PROBLEM.

*To find the equation of a straight line which shall be parallel to a given straight line.*

Let $$y = ax + b$$

be the equation of the given line.

The equation of the required line will be of the form

$$y = a'x + b',$$

in which $a'$ and $b'$ are undetermined.

The two right lines will be parallel, if they make the same angle with the axis of abscissas. Hence, if we make

$$a' = a$$

the second line will be parallel to the first; and its equation will be

$$y = ax + b',$$

in which equation $b'$ is undetermined, as it should be, since an infinite number of lines may be drawn parallel to a given line.

*Scholium.* If it be further required that the parallel shall pass through a given point, the position of the line will be entirely determined.

For, if the co-ordinate, of the given point be denoted by $x'$ and $y'$, and substituted in the last equation, we shall have

$$y' = ax' + b',$$

in which all the quantities are known excepting $b'$, which

is therefore determined in value: hence, the position of the line is fixed.

## PROPOSITION VII. PROBLEM.

*To find the angle included between two lines given by their equations.*

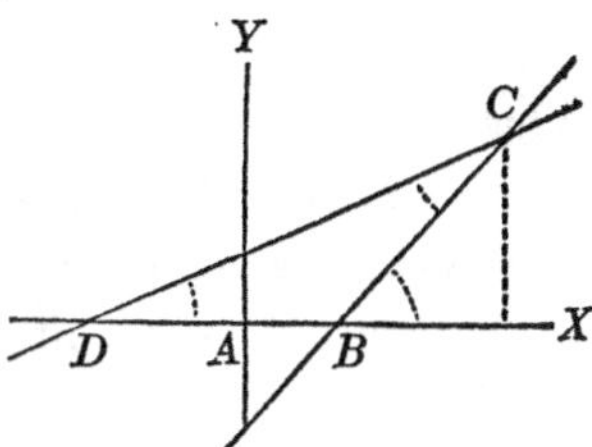


Let $DC$, $BC$, be the two given lines:

$y = ax + b$ the equation of the 1st.

$y = a'x + b'$ the equation of the 2d.

in which $a$, $a'$, $b$, $b'$, are known.

Denote the angles $CDX$ and $CBX$ by $\alpha$ and $\alpha'$, and the angle $DCB$ by $V$.

Then, since $\quad CBX = CDB + DCB,$

we have $\quad V = \alpha' - \alpha,$

and $\quad \text{tang } V = \text{tang } (\alpha' - \alpha) = \dfrac{\text{tang } \alpha' - \text{tang } \alpha}{1 + \text{tang } \alpha' \text{ tang } \alpha},$

to the radius of unity (Trig. Art. XXV).

Substituting for tang $\alpha'$, and tang $\alpha$, their values $a'$ and $a$, we have

$$\text{tang } V = \frac{a' - a}{1 + aa'}.$$

*Scholium* 1. If the lines become parallel, the angle $V$ will be 0, and hence,

$$\text{tang } V = \frac{a' - a}{1 + a'a} = 0 \qquad \text{(Trig. Art. VII).}$$

Therefore, $\quad a' - a = 0, \quad$ or $\quad a' = a,$

a relation already proved (Prop. VI).

*Scholium* 2. If the lines are perpendicular to each other, $V$ will be equal to 90°, and its tangent infinite (Trig. Art. IX): that is,

$$\text{tang } V = \frac{a' - a}{1 + a'a} = \infty;$$

hence, $$1 + a'a = 0 \quad \text{(Alg. Art. 109).}$$

This last is the equation of condition, by which two right lines are shown to be at right angles to each other. If one of the quantities, $a$, or $a'$, is known, the other can be found from the equation of condition.

### PROPOSITION VIII. PROBLEM.

*To determine the point in which two straight lines, given by their equations, intersect each other.*

Let $y = ax + b$ be the equation of the first line,

and $y = a'x + b'$, the equation of the second line.

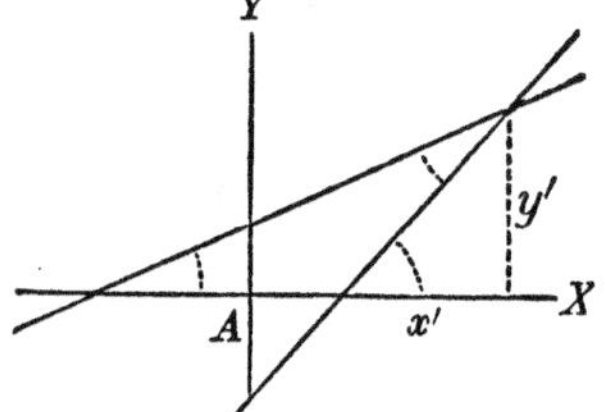


The point in which two straight lines intersect each other being found at the same time on both of the lines, its co-ordinates ought to satisfy both their equations.

If, therefore, we suppose $y$ and $x$ in the equation of the first line, to become equal to $y$ and $x$ in the equation of the second, the two equations will designate a point common to both the lines.

Combining the equations under this supposition, and designating the co-ordinates of the point of intersection by $x'$ and $y'$, we find,

$$x' = -\frac{(b - b')}{a - a'}, \quad \text{and} \quad y' = \frac{ab' - a'b}{a - a'}.$$

*Scholium* 1. If, in the two last equations, we suppose $a = a'$, the values of $x'$ and $y'$ will both become infinite, The supposition of $a = a'$ renders the two lines parallel, and therefore, their point of intersection ought to be at an infinite distance from both the co-ordinate axes.

If, at the same time, we also suppose $b = b'$, the values of $x'$ and $y'$ will become equal to 0 divided by 0, that is, indeterminate. But the two suppositions will cause the lines to coincide: hence, their point of intersection ought to be indeterminate, since every point of either line will satisfy both equations.

*Scholium* 2. The method which we have just employed for two straight lines, is general, and will serve to determine the points of intersection of curves whose equations are known.

For, the points whose co-ordinates will satisfy both the equations, must be common to the two curves. Hence, if we suppose the co-ordinates to be equal, and combine the equations under this supposition, the values of $x$ and $y$ found in the resulting equations will be the co-ordinates of points common to the two curves.

PROPOSITION IX. PROBLEM.

*To draw from a given point a line perpendicular to a given straight line, and to find the length of the perpendicular.*

Let $$y = ax + b,$$

be the equation of the given line, and $x'$, $y'$, the co-ordinates of the given point.

The equation of a straight line passing through the given point, will be of the form (Prop. IV),

$$y - y' = a' (x - x').$$

But since this line is to be perpendicular to the given line, we have (Prop. VII, sch. 2),

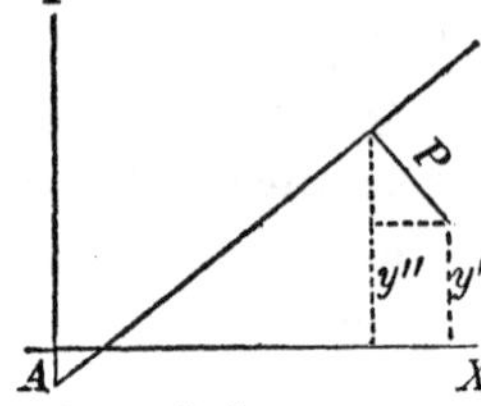


$$1 + aa' = 0,$$

from which we have,

$$a' = -\frac{1}{a}.$$

Substituting this value for $a'$, the equation of the perpendicular becomes,

$$y - y' = -\frac{1}{a}(x - x').$$

It is now required to find the length of the perpendicular. This is done by first finding the difference between the co-ordinates of the given point, and the co-ordinates of the point in which the perpendicular intersects the given line.

Let us designate the co-ordinates of this last point by $x''$, $y''$. Then, since the point is on the given line, its co-ordinates will satisfy the equation of the given line, and we shall have

$$y'' = ax'' + b\,;$$

and since the point is also on the perpendicular, its co-ordinates will also satisfy the equation of the perpendicular, and give

$$y'' - y' = -\frac{1}{a}(x'' - x').$$

If we eliminate $x''$ from these two equations, we shall have

$$y'' = \frac{a^2 y' + ax' + b}{1 + a^2}.$$

Subtracting $y'$ from both members, we obtain

$$y'' - y' = -\frac{y' - ax' - b}{1 + a^2}.$$

Substituting this value of $y''-y'$, in the last equation but one, and we have

$$x''-x'=+\frac{a(y'-ax'-b)}{1+a^2}.$$

Let us designate the length of the perpendicular by $P$. Since the distance between two points whose co-ordinates are $x'', y''$, $x', y'$, is (Prop. III),

$$\sqrt{(x''-x')^2+(y''-y')^2},$$

we have, by substituting for $x''-x'$, and $y''-y'$, their values found above,

$$P=\frac{y'-ax'-b}{\sqrt{1+a^2}}.$$

*Scholium.* If the given point should fall on the given line, its co-ordinates would satisfy the equation of the line, and give

$$y'=ax'+b.$$

This supposition would reduce the numerator of the value of $P$ to 0, and consequently $P$ would be equal to 0.

## *Transformation of Co-ordinates.*

The equations of a point determine its position with respect to the co-ordinate axes to which it is referred. The co-ordinate axes may be selected at pleasure, and the point may, at the same time, be referred to several systems.

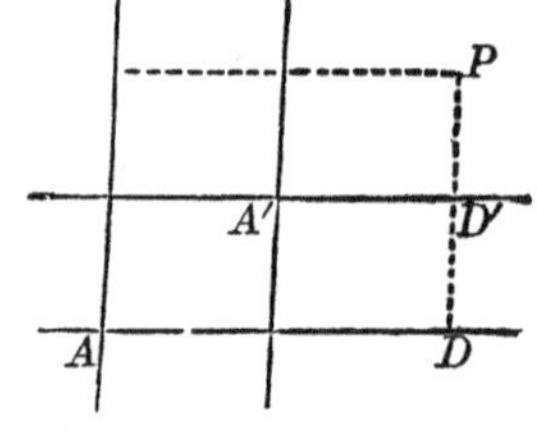


Let $A$, for example, be the origin of a system of co-ordinate axes, and $A'$ any point whose co-ordinates are $a$ and $b$.

Through $A'$ draw two new axes respectively parallel to the first.

The co-ordinates of any point, as $P$ referred to the first system, are $AD$, $PD$; and its co-ordinates referred to the second system are $A'D'$, $PD'$, and the point $P$ is equally determined to which ever system it be referred.

It is often necessary, for reasons that will be hereafter explained, to change the reference of points from one system of co-ordinate axes to another. This is called, the *transformation of co-ordinates.* The axes to which the points are first referred, are called the *primitive axes;* and the second axes to which they are referred, are called the *new axes.*

In changing the reference of points from one system of co-ordinate axes to another, it is necessary to find the co-ordinates of the points referred to the primitive axes, in terms of all the quantities in the new system on which they depend.

## PROPOSITION X. PROBLEM.

*To find the formulas for passing from one system of co-ordinate axes to another system, respectively parallel to the first.*

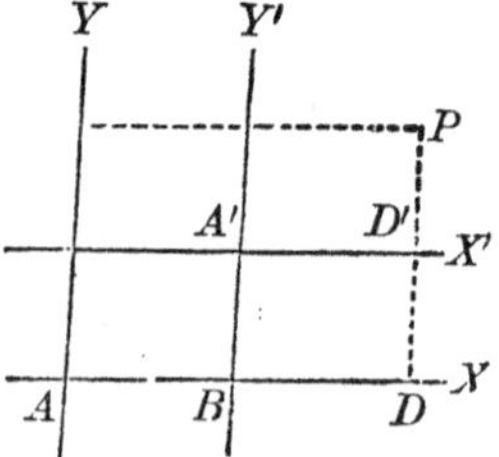


Let $A$ be the origin of the primitive system, and $A'$ the origin of the new system. Suppose the co-ordinates of the origin $A'$ to be $AB = a$, and $BA' = b$; and let us designate the co-ordinates of any point referred to the new axes by $x'$ and $y'$.

Then assuming any point, as $P$, we snaı have.

$$AD = AB + BD, \quad \text{and} \quad DP = DD' + D'P,$$

that is,

$$x = a + x', \quad \text{and} \quad y = b + y',$$

in which the primitive co-ordinates of any point are ex pressed in terms of the co-ordinates of the new origin ana the new co-ordinates of the same point.

*Scholium.* The new origin may be placed in either of the four angles of the primitive axes, by attributing proper signs to its co-ordinates $a$ and $b$. It is also to be observed, that $x'$ and $y'$ have the same algebraic signs in the different angles of the new system, as have been attributed to $x$ and $y$ in the corresponding angles of the primitive system.

PROPOSITION XI. PROBLEM.

*To find the formulas for passing from a system of rectangular, to a system of oblique co-ordinates, the origin remaining the same.*

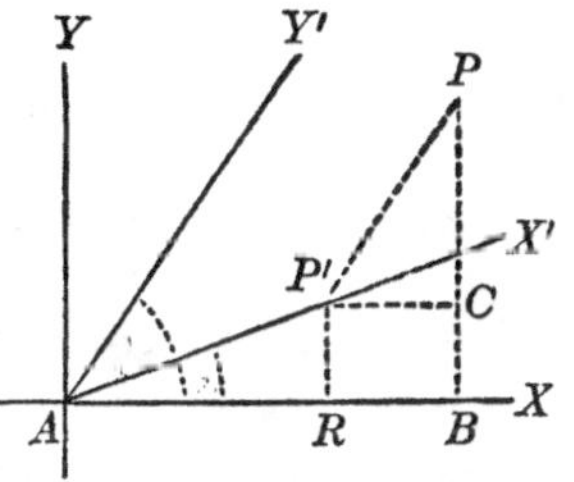


Let $A$ be the common origin, $AX$, $AY$, the primitive axes, and $AX'$, $AY'$, the new axes; and let us designate, as before, the co-ordinates of points referred to the new axes by $x'$ and $y'$.

Denote the angle which the new axis of $X'$ makes with the primitive axis of $X$ by $\alpha$, and the angle which $Y'$ makes with $AX$ by $\alpha'$, and let $P$ be any point in the plane of the axes. Through $P$, draw $PB$ parallel to the axis of $Y$, and $PP'$ parallel to the axis of $Y'$. draw also $P'R$ parallel to $Y$, and $P'C$ parallel to the axis of $X$

Then, $\quad AB = AR + RB$

will be the abscissa of $P$, referred to the primitive axes;

and $\quad PB = BC + CP,$

will be its ordinate.

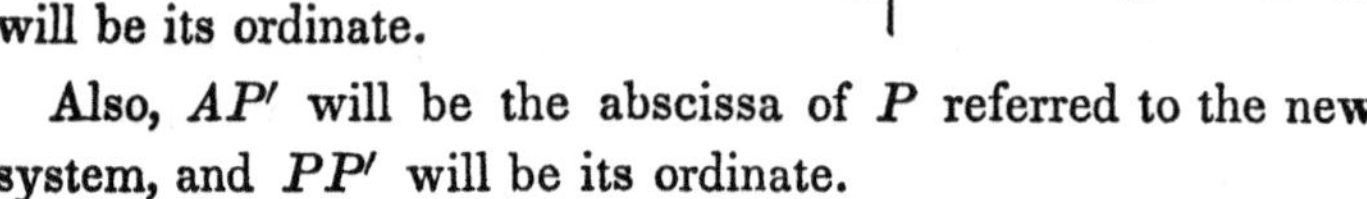


Also, $AP'$ will be the abscissa of $P$ referred to the new system, and $PP'$ will be its ordinate.

But $\quad AR = AP' \cos \alpha \quad$ (Trig. Th. I, Cor.),

that is, $\quad AR = x' \cos \alpha,$

and $\quad RB = P'C = PP' \cos \alpha' = y' \cos \alpha' \cdot$

hence, $\quad x = x' \cos \alpha + y' \cos \alpha'.$

We also have, $\quad P'R = CB = AP' \sin \alpha \quad$ (Trig. Th. I, Cor.),

that is, $\quad CB = x' \sin \alpha,$

and $\quad PC = PP' \sin \alpha' = y' \sin \alpha' :$

hence, $\quad y = x' \sin \alpha + y' \sin \alpha'.$

Hence, the formulas are,

$$x = x' \cos \alpha + y' \cos \alpha', \qquad y = x' \sin \alpha + y' \sin \alpha'.$$

*Scholium.* If it were required, at the same time, to change the origin to a point whose co-ordinates, when referred to the primitive system, are $a$ and $b$, the formulas would become,

$$x = a + x' \cos \alpha + y' \cos \alpha', \qquad y = b + x' \sin \alpha + y' \sin \alpha'$$

### PROPOSITION XII. PROBLEM.

*To find the formulas for passing from a system of oblique co-ordinates to a system of rectangular co-ordinates, the origin remaining the same.*

Take the formulas of the last problem

$$x = x' \cos \alpha + y' \cos \alpha', \quad y = x' \sin \alpha + y' \sin \alpha'$$

If now, we regard the oblique as the primitive axes, we must find the co-ordinates of points referred to these axes in terms of the rectangular co-ordinates and the angles $\alpha$ and $\alpha'$.

If we multiply the first equation by the sin $\alpha'$, and the second by cos $\alpha'$, and then subtract them, $y'$ will be eliminated; and if $x'$ be eliminated in a similar manner, we shall obtain

$$x' = \frac{x \sin \alpha' - y \cos \alpha'}{\sin(\alpha' - \alpha)}, \quad y' = \frac{y \cos \alpha - x \sin \alpha}{\sin(\alpha' - \alpha)}.$$

*Scholium.* If the origin were changed, at the same time, to a point whose co-ordinates, with reference to the oblique system, are $a$ and $b$, we should have,

$$x' = a + \frac{x \sin \alpha' - y \cos \alpha'}{\sin(\alpha' - \alpha)} \quad y' = b + \frac{y \cos \alpha - x \sin \alpha}{\sin(\alpha' - \alpha)}.$$

### PROPOSITION XIII. PROBLEM.

*To find the formulas for passing from a system of rectangular co-ordinates to a system of co-ordinates also rectangular, the origin remaining the same.*

If in the equations of (Prop. XI), which are

$$x = x' \cos \alpha + y' \cos \alpha', \quad y = x' \sin \alpha + y' \sin \alpha',$$

we make

$$\alpha' - \alpha = 90°,$$

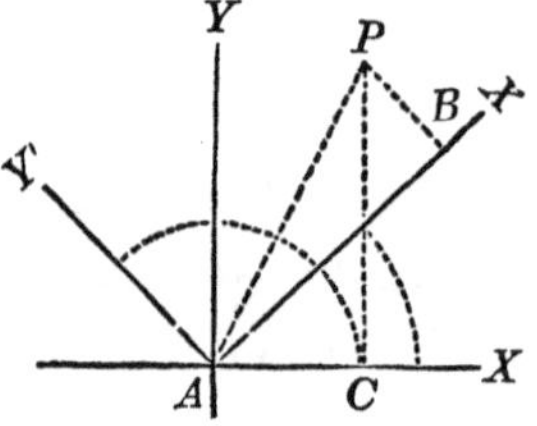


the new system of co-ordinates, as well as the primitive, will be at right angles.

This supposition will give

$$\alpha' = 90° + \alpha. \quad \text{and}$$

$$\sin \alpha' = \sin 90° \cos \alpha + \cos 90° \sin \alpha = + \cos \alpha$$

$$\cos \alpha' = \cos 90° \cos \alpha - \sin 90° \sin \alpha = - \sin \alpha.$$

Substituting these values in the first equations, and they become

$$x = x' \cos \alpha - y' \sin \alpha, \qquad y = x' \sin \alpha + y' \cos \alpha.$$

*Scholium* 1. If the origin be changed at the same time, the equations will become,

$$x = a + x' \cos \alpha - y' \sin \alpha, \qquad y = b + x' \sin \alpha + y' \cos \alpha.$$

*Scholium* 2. If we square the two equations first found, and then add them together, member by member, observing, that the sine square of an arc plus the cosine square, is equal to the square of the radius, or 1, we shall find

$$x^2 + y^2 = x'^2 + y'^2.$$

This is as it should be, since $x^2 + y^2$ or $\overline{AC}^2 + \overline{CP}^2$ is equal to $\overline{AP}^2$: and since the new system is also rectangular, $x'^2 + y'^2$ or $\overline{AB}^2 + \overline{PB}^2$, is likewise equal to $\overline{AP}^2$.

PROPOSITION XIV. PROBLEM.

*To find the formulas for passing from a system of oblique co-ordinates to a new system of co-ordinates also oblique—the origin remaining the same.*

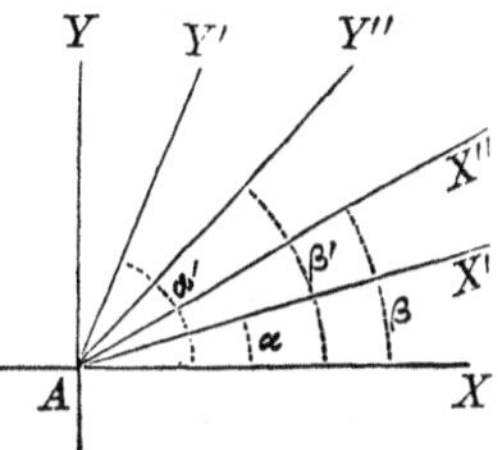


Let $AX'$ and $AY'$ be the primitive axes, $AX''$, $AY''$, the new axes, and $A$ the common origin. Through $A$ draw the two axes $AX$, $AY$, at right angles to each other. Designate the co-ordinates of points referred to the primitive system by $x'$ and $y'$, those referred to the new system by $x''$ and $y''$, and the co-ordinates of the rectangular system, by $x$ and $y$; and designate the angles as marked in the figure.

Now the formulas to pass from the rectangular to the first oblique system are (Prop. XI),

$$x = x' \cos \alpha + y' \cos \alpha', \qquad y = x' \sin \alpha + y' \sin \alpha'$$

The formulas for passing from the rectangular to the second oblique system, are

$$x = x'' \cos \beta + y'' \cos \beta', \qquad y = x'' \sin \beta + y'' \sin \beta'.$$

If we eliminate $x$ and $y$ from these equations, the resulting equations will express the relations which exist between the primitive co-ordinates $x'$, $y'$, and the new co-ordinates $x''$, $y''$. These relations are

$$x' \cos \alpha + y' \cos \alpha' = x'' \cos \beta + y'' \cos \beta'.$$

$$x' \sin \alpha + y' \sin \alpha' = x'' \sin \beta + y'' \sin \beta'.$$

If from these equations we first eliminate $y'$ and then $x'$, we shall find

$$x' = \frac{x'' \sin(\alpha' - \beta) + y'' \sin(\alpha' - \beta')}{\sin(\alpha' - \alpha)},$$

$$y' = \frac{x'' \sin(\beta - \alpha) + y'' \sin(\beta' - \alpha)}{\sin(\alpha' - \alpha)}.$$

in which the angles formed by the rectangular axes do not enter, since

$$\alpha' - \beta = Y'AX'', \qquad \alpha' - \beta' = Y'AY'', \qquad \beta - \alpha = X''AX',$$

$$\beta' - \alpha = Y''AX', \qquad \alpha' - \alpha = Y'AX'.$$

*Scholium* 1. If the origin of co-ordinates be changed at the same time, the equations will become,

$$x' = a + \frac{x'' \sin(\alpha' - \beta) + y'' \sin(\alpha' - \beta')}{\sin(\alpha' - \alpha)},$$

$$y' = b + \frac{x'' \sin(\beta - \alpha) + y'' \sin(\beta' - \alpha)}{\sin(\alpha' - \alpha)}.$$

*Scholium* 2. The primitive co-ordinates of any point determined with reference to a new system, depend for their values,

1st. On the position of the new origin.

2d. On the angles which the new axes make with the primitive axes, and

3d. On the co-ordinates of the same point referred to the new system.

*Scholium* 3. The transformation of co-ordinates embraces two distinct classes of propositions.

1st. To transfer the reference of points from one system of co-ordinate axes to another system which is known. In this case the co-ordinates of the new origin and the angles which the new axes make with the primitive axes, are known.

2d. It is often required, however, so to dispose of the

new origin, and to give such directions to the new axes, as to cause the resulting equations to fulfil certain conditions, or to assume a certain form. In this case, the conditions imposed determine the position of the new origin and the directions of the new axes.

*Scholium* 4. Since the primitive co-ordinates are always determined in linear functions of the new co-ordinates, that is, by equations of the first degree, the substitution of their values in the equation of any line, will not alter the degree of that equation. Hence, *the given equation of a line, and its equation when referred to a new system of co-ordinate axes, will always be of the same degree.*

*Scholium* 5. We shall terminate this subject by a single example.

Having given the equation of a straight line,

$$y = a'x + b',$$

referred to rectangular co-ordinates, it is required to find its equation when the line is referred to oblique co-ordinates having a different origin. We have (Prop. XI, sch.),

$$x = a + x' \cos \alpha + y' \cos \alpha' \quad y = b + x' \sin \alpha + y' \sin \alpha'.$$

Substituting these values for $x$ and $y$ in the equation of the line, we have,

$$b + x' \sin \alpha + y' \sin \alpha' = a'(a + x' \cos \alpha + y' \cos \alpha') + b':$$

or, by reducing

$$y' = x' \frac{a' \cos \alpha - \sin \alpha}{\sin \alpha' - a' \cos \alpha'} + \frac{aa' + b' - b}{\sin \alpha' - a' \cos \alpha'},$$

which is the equation of the straight line, referred to the oblique axes.

## *Of Polar Co-ordinates.*

We have seen, that the relative position of points and lines may be determined analytically, by referring them to rectilinear axes, which make with each other a given angle. There are also other methods by which they may likewise be determined.

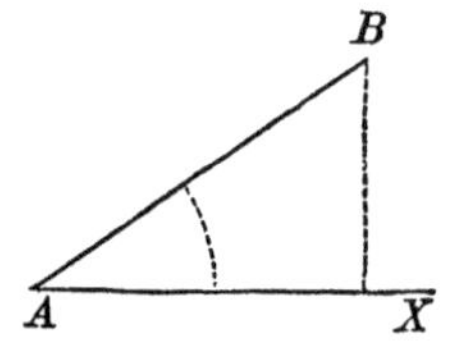


Assume, for example, any point, as $A$, and through it draw any straight line as $AX$. If we suppose a straight line, as $AB$, to be turned around the point $A$, so as to make with $AX$ all possible angles, from 0 to 360°, and suppose at the same time the line $AB$ to increase or diminish at pleasure, the extremity $B$ may be made to coincide, in succession, with every point of the plane.

There are two variable quantities to be considered: 1st. the variable angle $XAB$; and 2dly, the variable distance $AB$; and every point in the plane may be determined by attributing suitable values to these variables.

This method of determining points by a variable angle and a variable distance, is called the *system of polar co-ordinates*. The variable distance $AB$, is called the *radius-vector*; and the fixed point $A$, from which it is estimated, is called the *pole*.

Designate the variable angle $XAB$, by $v$, the radius-vector $AB$, by $r$, and the co-ordinates of the point $B$, referred to rectangular axes, by $x$ and $y$; then, if the origin of the rectangular axes be at $A$, we shall have,

$$x = r \cos v \quad \text{(Trig. Th. I, Cor.)},$$

and $$y = r \sin v \quad \text{(Trig. Th. I, Cor.)}.$$

From the first equation, we have

$$r = \frac{x}{\cos v}.$$

Now, since $x$ and the $\cos v$ are both positive in the first and fourth angles, and both negative in the second and third, they will always be affected with the same sign, and hence the sign of $r$ will be constantly positive: consequently, *a negative value of the radius-vector can never enter into the analysis.*

If therefore, such a value should be obtained, we ought to infer, that incompatible conditions have been introduced into the equations; and hence, *all negative values of the radius-vector must be rejected.*

PROPOSITION XV. PROBLEM.

*To find the formulas for passing from a system of rectangular to a system of polar co-ordinates.*

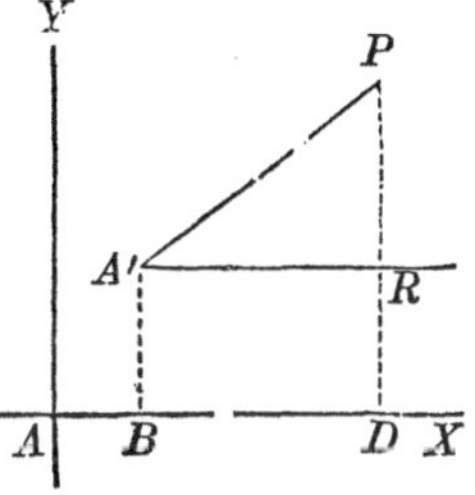


Let $A$ be the origin of the co-ordinate axes, $A'$ the pole, $A'R$ parallel to $AX$, the line from which the variable angles are estimated, and $A'P$ the radius-vector of the point $P$. Let the co-ordinates of the pole $A'$ be represented by $a$ and $b$.

Now, $A'R = r \cos v$, and $PR = r \sin v$.

But $AD = AB + BD$

and $PD = DR + PR$;

hence, $x = a + r \cos v$,

and $y = b + r \sin v$;

which are the formulas required

*Scholium* 1. If the pole $A'$ be placed at the origin $A$, the equations will become,

$$x = r\cos v, \qquad y = r\sin v.$$

*Scholium* 2. If, instead of estimating the variable angle $v$ from the line $A'R$, parallel to $AX$, it be estimated from $A'R'$, which makes with $AX$ a given angle $\alpha$, the equations will become

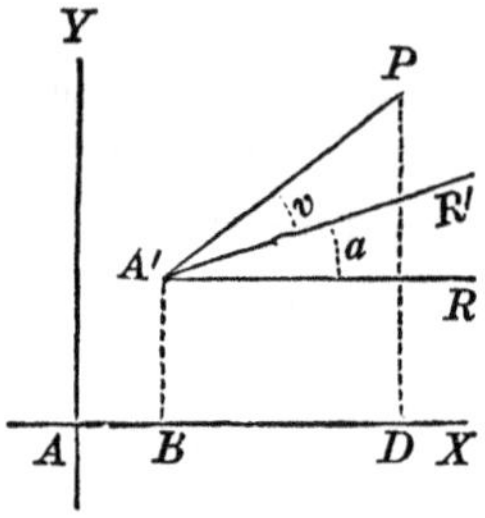

$$x = a + r\cos(v + \alpha),$$

$$y = b + r\sin(v + \alpha).$$

# BOOK III.

## *Of the Circle.*

The equation of a line expresses the relation which exists between the co-ordinates of every point of the line (Bk. II, Art. 3). All lines in which this relation can be expressed algebraically, that is, by algebraic quantities alone, are called *algebraic lines,* and these are the only lines which will be considered in Analytical Geometry.

Lines are divided into different orders, according to the degree of their equations. For example, the right line is a line of the first order, since its equation is of the first degree. The circumference of the circle is a line of the second order, its equation being of the second degree (Bk. II, Art. 2); and if the equation of a line were of the third degree, the line would be of the third order.

The *discussion of an equation,* consists in classing the line which the equation represents; in determining its position, its form, its limits, and the points in which it intersects the co-ordinate axes.

### PROPOSITION I. PROBLEM.

*To find the equation of the circumference of a circle, and to discuss it.*

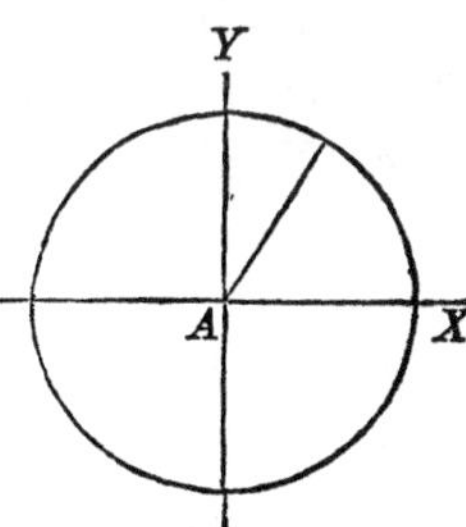


Let $A$ be the origin, and $AX$, $AY$, the co-ordinate axes.

It is required to find the equation of a curve such, that all its points shall be at a given distance from the origin $A$. Let $R$ designate that distance, and $x$ and $y$, the co-ordinates of any point of the curve. The square of the distance from the origin to any point, whose co-ordinates are $x$ and $y$, (Bk. II, Prop. III, Sch.), is

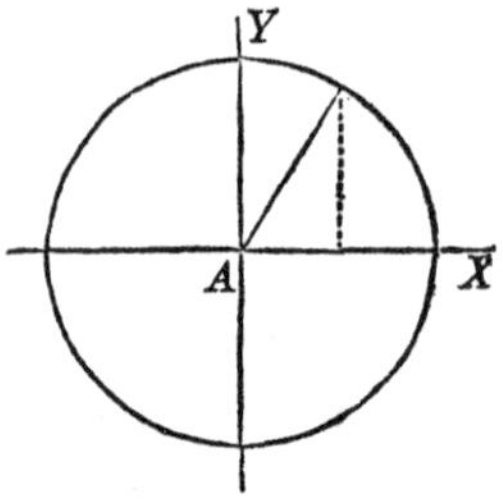

$$x^2 + y^2;$$

hence,
$$x^2 + y^2 = R^2,$$

which is the equation required.

In discussing the equation, we begin by determining the points in which the circumference cuts the co-ordinate axes.

The co-ordinates of these points must satisfy, at the same time, both the equation of the circle, and the equations of the axes.

The equations of the axis of $X$ being

$$y = 0 \quad \text{and} \quad x \quad \text{indeterminate};$$

if we make $y = 0$ in the equation of the circle, the corres ponding values of $x$ will be the abscissas of those points which are common to the circumference and the axis of $X$. that is,

$$x = \pm R;$$

which shows that the curve cuts the axis of abscissas in two points, one on each side of the origin, and both at a distance from it equal to the radius of the circle.

To find the points in which the circumference cuts the axis of $Y$, make $x = 0$, and there results,

$$y = \pm R;$$

the axis of $Y$, therefore, intersects the circumference in two points, equally distant from the origin, one above the axis of $X$, and the other below it.

To trace the curve through the intermediate points, find the value of $y$ from the equation, which gives,

$$y = \pm \sqrt{R^2 - x^2}.$$

Now, since every value for $x$ gives for $y$ two equal values, with contrary signs, it follows that the curve is symmetrical with respect to the axis of $X$: and in the same manner it might be shown to be symmetrical with respect to the axis of $Y$.

From $x = 0$, which gives

$$y = \pm R,$$

the values of $y$ decrease, as $x$ increases: and when $x$ be comes equal to $\pm R$, we have

$$y = 0.$$

If $x$ becomes greater than $\pm R$, the values of $y$ become imaginary, which shows that the curve is limited both in the direction of $x$ positive, and of $x$ negative.

By placing the equation under the form

$$x = \pm \sqrt{R^2 - y^2},$$

we may show that the circumference is also limited in the direction of $y$ positive, and in that of $y$ negative.

By attributing a particular value to either of the variables in the equation

$$y = \pm \sqrt{R^2 - x^2},$$

the corresponding value of the other variable may be found from the equation

If we suppose $R = 1$, and then make

| | | |
|---|---|---|
| $x=0$ | we have | $y=\pm 1,$ |
| $x=\frac{1}{2}$ | gives | $y=\pm\sqrt{\frac{3}{4}}=\frac{1}{2}\sqrt{3},$ |
| $x=\frac{3}{4}$ | gives | $y=\pm\sqrt{\frac{7}{16}}=\frac{1}{4}\sqrt{7},$ |
| &c. | &c. | &c. &c. |

*Scholium* 1. If in the equation

$$x^2+y^2=R^2,$$

$x$ and $y$ be taken to represent the co-ordinates of a point within the circumference, the equality will be destroyed, and $x^2+y^2$ will be less than $R^2$, and we shall have

$$x^2+y^2-R^2<0,$$

that is, negative.

For a point on the curve

$$x^2+y^2-R^2=0,$$

and for a point without the curve

$$x^2+y^2-R^2>0,$$

is positive.

*Scholium* 2. The equation

$$y^2=R^2-x^2,$$

may be put under the form

$$y^2=(R+x)(R-x),$$

in which the factors, $R+x$ and $R-x$ are the two segments into which the ordinate $y$ divides the diameter: this ordinate is, therefore, a mean proportional between the two segments.

*Scholium* 3. The equation of the circle may also be placed under another form, by transferring the origin of co-ordinates from the centre to a point of the circumference.

If in this transformation the new co-ordinate axes be parallel to the primitive axes, we shall have the formulas (Bk. II, Prop. X),

$$x = a + x' \quad \text{and} \quad y = b + y'.$$

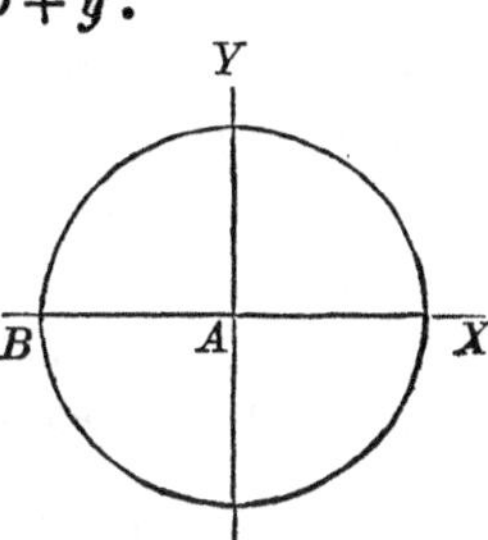


Let it be required to transfer the origin to $B$.

The co-ordinates of this point, are

$$a = -R \quad \text{and} \quad b = 0;$$

hence, $x = x' - R$ and $y = y'$.

Substituting these values in the equation

$$y^2 = R^2 - x^2$$

we obtain

$$y'^2 = 2Rx' - x'^2,$$

or omitting the accents

$$y^2 = 2Rx - x^2;$$

which is the equation of the circle when the origin of co-ordinates is in the circumference.

*Scholium* 4. There is yet a more general form under which the equation of a circle may be expressed.

The characteristic property of the circumference of a circle is, that all the points are at an equal distance from the centre. To express this property *analytically*, and in a general manner, designate the co-ordinates of the centre by $x'$ and $y'$, the co-ordinates of any point of the circumference by $x$ and $y$, and the radius by $R$.

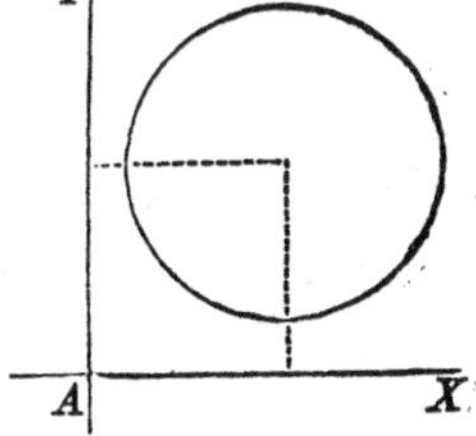


Now the distance from any point whose co-ordinates are $x', y'$ to a point whose co-ordinates are $x$ and $y$, is (Bk. II, Prop. III),

$$(x - x')^2 + (y - y')^2 = R^2.$$

This is, therefore, the most general equation of the circle referred to rectangular co-ordinates. By attributing proper values and signs to $x'$ and $y'$, the centre of the circle may be placed at any point in the plane of the co-ordinate axes.

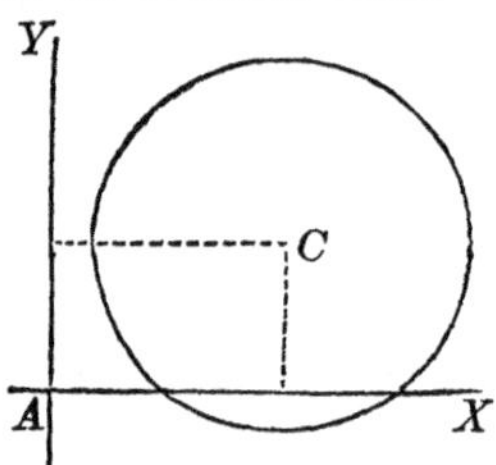


To find the points in which the circumference intersects the axis of $X$, make $y = 0$, and we have

$$x = x' \pm \sqrt{R^2 - y'^2},$$

in which we see that the values of $x$ will become imaginary when $y'$ exceeds $R$, and it is plain that in that case there would not be an intersection.

To find the points in which the circumference intersects the axis of $Y$, make $x = 0$, and we have

$$y = y' \pm \sqrt{R^2 - x'^2},$$

in which the value of $y$ will be imaginary, if $x'$ exceeds $R$.

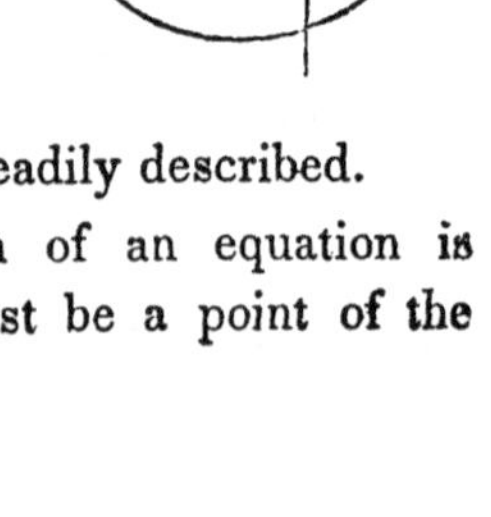

If the co-ordinates of the centre of a circle are

$$x' = -2 \quad \text{and} \quad y' = -4,$$

and the radius equal to 6, its equation will be

$$(x + 2)^2 + (y + 4)^2 = 36,$$

from which the circumference may be readily described.

*Scholium* 5. If the absolute term of an equation is wanting, the origin of co-ordinates must be a point of the curve.

For, if $x$ and $y$ be each made equal to 0, all the terms will reduce to 0 and, therefore, the equation will be satisfied. But,

$$x=0 \quad \text{and} \quad y=0,$$

are the equations of the origin, and since the co-ordinates ot the origin satisfy the equation of the curve, the origin must be a point of the curve.

Two lines which are drawn through the two extremities of any diameter of a curve, and which intersect the curve at the same point, are called *supplementary* chords.

PROPOSITION II. THEOREM.

*The supplementary chords in the circle are perpendicular to each other.*

Let $A$ be the origin of co-ordinates, and $B$ and $B'$ the extremities of a diameter.

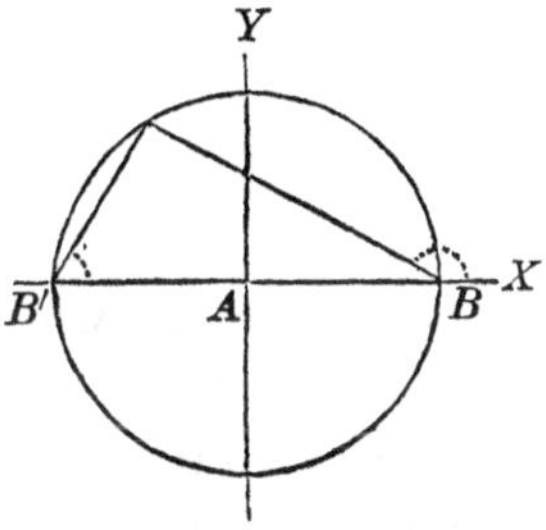

The equation of a straight line passing through a given point is of the form (Bk. II, Prop. IV)

$$y-y'=a(x-x').$$

If the line be made to pass through $B$, whose co-ordinates are $y'=0$, and $x'=+R$, its equation will become

$$y=a(x-R).$$

For a like reason, the equation of a straight line passing through $B'$, is

$$y=a'(x+R).$$

If these two lines intersect each other, the co-ordinates of their point of intersection will satisfy both equations. Hence, if we suppose $x$ in one equation to be equal to $x$ in the other, and $y$ equal to $y$, and then combine the equations by multiplying them together, the resulting equation,

$$y^2 = aa'(x^2 - R^2),$$

will express the condition that the two straight lines shall intersect on the plane of the co-ordinate axes.

But if the point of intersection is to be in the circumference of the circle, $x$ and $y$ must satisfy the equation

$$x^2 + y^2 = R^2,$$

or $$y^2 = R^2 - x^2 = -1(x^2 - R^2).$$

Hence, $$aa' = -1, \quad \text{or} \quad aa' + 1 = 0.$$

The two lines are therefore perpendicular to each other (Bk. II, Prop. VII, Sch. 2).

*Scholium* 1. In the equation of condition,

$$aa' + 1 = 0,$$

the two tangents $a$ and $a'$ are undetermined; there are, therefore, an infinite number of values which may be attributed to them that will satisfy the equation; which shows that there are an indefinite number of supplementary chords that may be drawn through the extremities of the same diameter, each pair of which will be perpendicular to each other.

*Scholium* 2. If it be required, that one of the supplementary chords shall make a given angle with the axis of $X$, its tangent $a$ or $a'$ becomes known, and then the value of the other tangent may be found from the equation of condition

If either $a$ or $a'$ is equal to 0, the other will be infinite; which shows, that if one of the chords coincides with the axis of $X$, the other will become perpendicular to it.

PROPOSITION III. PROBLEM.

*To find the equation of a straight line which shall be tangent to the circumference of a circle.*

Let $A$ be the origin of co-ordinates,

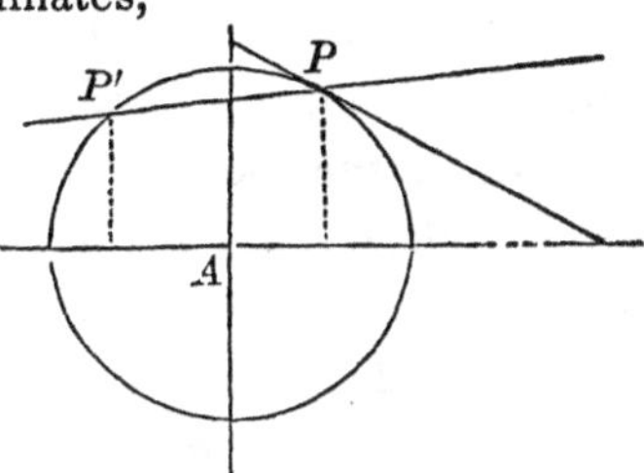


and $$x^2 + y^2 = R^2,$$

the equation of the circle.

Take any point of the circumference, as $P$, and designate its co-ordinates by $x'', y''$. Through this point draw a secant line; its equation will be of the form

$$y - y'' = a(x - x''):$$

it is now required to find the value of $a$ when the secant line $PP'$ becomes tangent to the circumference.

Since the point $P$ is in the circumference, its co-ordinates will satisfy the equation of the circle, and we shall have,

$$x''^2 + y''^2 = R^2.$$

Subtracting this from the equation of the circle, we obtain

$$x^2 - x''^2 + y^2 - y''^2 = 0,$$

or $$(x + x'')(x - x'') + (y + y'')(y - y'') = 0;$$

in which equation, $x$ and $y$ are the co-ordinates of any point of the circumference.

If this equation be combined with the equation of the secant, $x$ and $y$ in the resulting equation, will be the co-ordinates of $P'$, the second point in which the secant inter sects the circumference. The equations are most readily combined by substituting for $y - y''$, the value found in the equation of the secant. Making the substitution, we obtain

$$(x+x'')(x-x'')+(y+y'')a(x-x'')=0,$$

and dividing by $x-x''$, we have

$$x+x''+a(y+y'')=0.$$

If now we suppose the secant $PP'$ to turn around the point $P$, the point $P'$ will approach $P$: and when $P'$ shall coincide with $P$, the secant line will become tangent to the circumference. When this takes place, we shall have

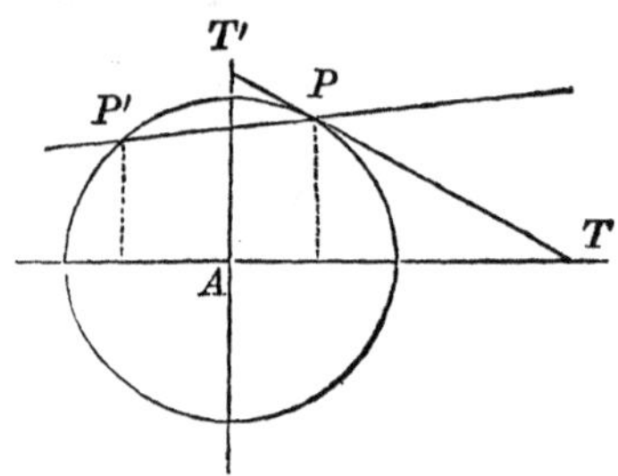


$$x=x'' \qquad \text{and} \qquad y=y'',$$

and the last equation will give,

$$a=-\frac{x''}{y''},$$

in which $x''$, $y''$, are the co-ordinates of the point of contact.

Substituting this value in the equation of the line passing through $P$, and we have,

$$y-y''=-\frac{x''}{y''}(x-x'');$$

or, by reducing

$$yy''-y''^2=-x''x+x''^2,$$

or $$yy''+xx''=y''^2+x''^2,$$

or $$yy''+xx''=R^2;$$

in which $x$ and $y$ are the general co-ordinates of the tangent line.

*Scholium* 1. For the point in which the tangent intersects the axis of $Y$, we have $x=0$, and

$$y = \frac{R^2}{y''} = AT'.$$

And for the point in which it intersects the axis of $X$ $y = 0$, and

$$x = \frac{R^2}{x''} = AT.$$

*Scholium* 2. We can readily prove that every point of the tangent line, except the point of contact, lies without the circumference of the circle.

For, we have the equations

$$yy'' + xx'' = R^2,$$

$$y''^2 + x''^2 = R^2:$$

and if we subtract twice the first equation from the second, there will result,

$$y''^2 - 2yy'' + x''^2 - 2xx'' = -R^2.$$

Adding $x^2 + y^2$ to each member, the result may be put under the form

$$(y - y'')^2 + (x - x'')^2 = x^2 + y^2 - R^2.$$

In this equation $x$ and $y$ are the general co-ordinates of the tangent line. But the first member is always positive, being the sum of two squares: hence, the second member is positive; and therefore all the points of the tangent line lie without the circumference (Prop. I, Sch. 1).

If we consider the point of the tangent line at which it touches the circle, we shall have

$$y = y'', \qquad \text{and} \qquad x = x''.$$

Each member of the equation will then become equal to 0, which shows, that the co-ordinates of this point will satisfy the equation of the circle

*Scholium* 3. We have thus far considered the tangent line as touching the circumference at a given point. Let us now examine the conditions which are to be fulfilled when the tangent line is required to pass through a given point without the circle.

The equation of the tangent line is of the form

$$yy'' + xx'' = R^2.$$

If this line be required to pass through a given point whose co-ordinates are $x'$, $y'$, the equation will become

$$y'y'' + x'x'' = R^2.$$

Subtracting this from the first equation, and we have

$$(y - y')y'' + (x - x')x'' = 0,$$

or
$$y - y' = -\frac{x''}{y''}(x - x'),$$

for the equation of a tangent line drawn through a given point without the circle.

If it be required to find the values of the co-ordinates $x''$, $y''$, of the point of contact, in terms of known quantities, we have the two equations,

$$y'y'' + x'x'' = R^2,$$

$$y''^2 + x''^2 = R^2,$$

from which the values of $x''$, $y''$, may be found in terms of the known quantities, $x'$, $y'$, and $R$.

Substituting the values thus found in the equation of the tangent line passing through a given point, and that line will become entirely determined, and will satisfy all the conditions.

The last equation being of the second degree, will give two values for $x''$, $y''$; and therefore, two tangent lines may be drawn to the circle from a given point without.

Instead, however, of finding the values of $x''$ and $y''$, it is better to place the equations under such a form as will indicate the geometrical construction by which the tangent line is to be drawn.

If we subtract the first equation from the second, we find,

$$y''^2 - y'y'' + x''^2 - x'x'' = 0.$$

If we add $\frac{x'^2}{4}$ and $\frac{y'^2}{4}$ to both members, the result may be placed under the form

$$\left(y'' - \frac{y'}{2}\right)^2 + \left(x'' - \frac{x'}{2}\right)^2 = \frac{x'^2 + y'^2}{4}.$$

By comparing this with the most general equation of the circle (Prop. I, Sch. 4), we see that the point of tangency whose co-ordinates are $x''$, $y''$, is in the circumference of a circle the co-ordinates of whose centre are $\frac{y'}{2}$ and $\frac{x'}{2}$, and whose radius is $\frac{1}{2}\sqrt{x'^2 + y'^2}$, and this circumference has for its diameter $\sqrt{x'^2 + y'^2}$.

But the point of contact is also found on the circumference whose equation is

$$x''^2 + y''^2 = R^2.$$

Since, therefore, the point of contact is found at the same time in the circumferences of two circles, it must be found at their points of intersection.

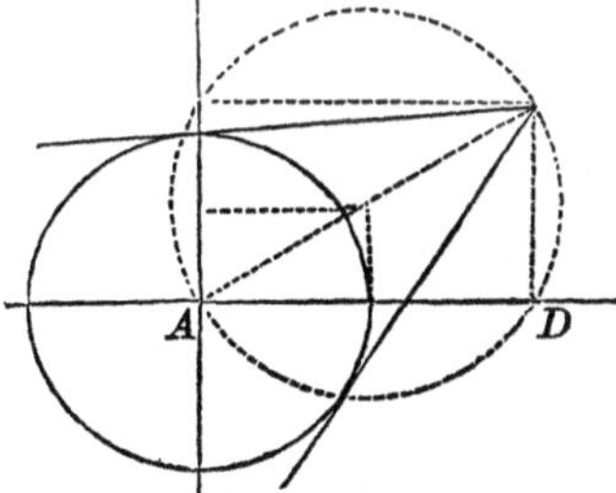


To make the construction for the tangent, we lay off from $A$, the origin of co-ordinates, the abscissa $x' = AD$, and at the point so determined, erect a perpendicular to the axis of $X$ and make it equal to $y'$. Join the extremity of the perpendicular and the origin $A$, and on this line as a diameter describe a circumference; the points $P$ and $P'$ in which it intersects the circumference of the given circle will be the points of tangency. There will be two tangents, since the circumferences will intersect each other in two points.

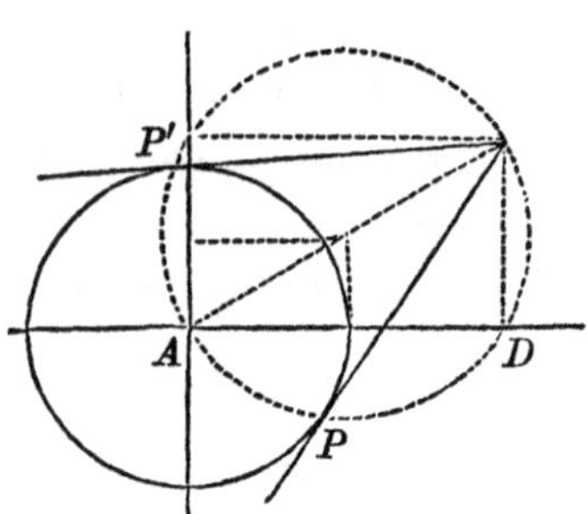


The construction will also show when the problem is impossible. For, if the values of $x'$, $y'$, are such, that the point falls within the circle, the two circumferences will not intersect each other.

A line is said to be *normal* to a curve when it is perpendicular to the tangent line at the point of contact. A line so drawn, is called a *normal line.*

### PROPOSITION IV. THEOREM.

*Every normal line in a circle passes through the centre.*

The equation of a straight line passing through the point of tangency will be of the form

$$y - y'' = a'(x - x'').$$

The condition of its being perpendicular to the tangent will give

$$aa' + 1 = 0, \quad \text{or} \quad a' = -\frac{1}{a}.$$

But from the equation of the tangent, we have

$$a = -\frac{x''}{y''},$$

hence, $$a' = \frac{y''}{x''}.$$

The equation of the normal therefore becomes

$$y - y'' = \frac{y''}{x''}(x - x''),$$

or by reducing

$$yx'' - y''x = 0,$$

and since this equation has no absolute term, the line passes through the origin of co-ordinates. We have thus proved a property known in elementary geometry, viz.: that a tangent line is perpendicular to the radius which passes through the point of contact.

## *Of the Polar Equation of the Circle.*

The polar equation of a curve, is the equation which is obtained by referring the curve to a fixed point and a given right line. The fixed point is called *the pole;* the variable distance from the pole to any point of the curve is called the *radius-vector*, and the angle which the radius-vector makes with the given straight line, is called the *variable angle.*

### PROPOSITION V. PROBLEM.

*To find the polar equation of the circle.*

The equation of the circle referred to rectangular co-ordinates, having their origin at the centre, is

$$x^2 + y^2 = R^2.$$

It is required to find the equation of the circle referred to a system of polar co-ordinates.

Let $A$ be the origin of the co-ordinate axes, $P$ the position of the pole, and $PX'$ parallel to the axis of $X$, the line from which the variable angle is estimated. Denoting the co-ordinates of the pole by $a$ and $b$, the radius-vector by $r$, and the variable angle by $v$, we shall have for passing from a system of rectangular to a system of polar co-ordinates (Bk. II, Prop. XV), the following formulas :

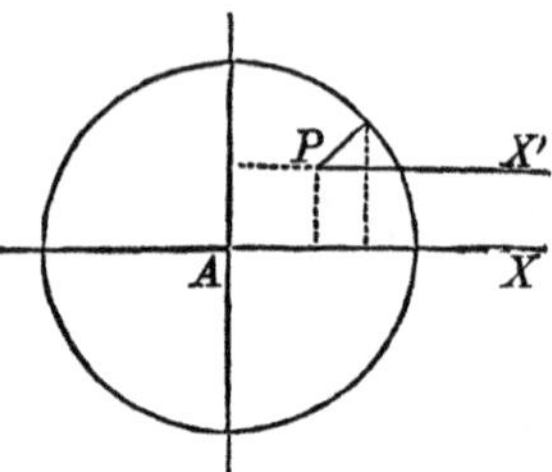


$$x = a + r\cos v, \qquad y = b + r\sin v.$$

Squaring each member of the two equations and substituting the values of $x^2$, $y^2$, thus found, in the equation of the circle, and recollecting that $\cos^2 v + \sin^2 v = 1$, we shall obtain,

$$r^2 + 2(a\cos v + b\sin v)r + a^2 + b^2 - R^2 = 0,$$

which is the polar equation of the circle.

*Scholium* 1. The pole $P$ may be placed at any point in the plane of the co-ordinate axes, by attributing suitable values and signs to its co-ordinates $a$ and $b$.

Let it be required, for example, to place the pole at $B'$, and to discuss the equation. The co-ordinates of the point $B'$ are,

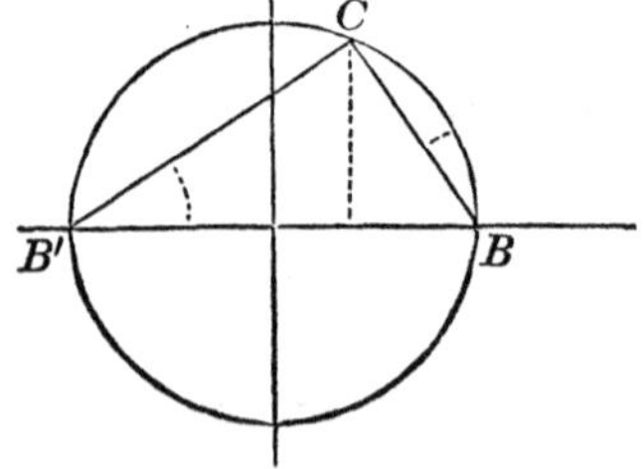


$$x = -R \quad \text{and} \quad y = 0;$$

hence, $\quad a = -R \quad$ and $\quad b = 0.$

Therefore, when the pole is placed at $B'$, the polar equation becomes,

$$r^2 - 2R\cos v\, r = 0,$$

in which the angle $v$ is estimated from the axis of $X$.

Now, since in this equation the absolute term is wanting, one of the roots is equal to 0 (Alg. Art. 148); which ought to be the case, since the pole is on the curve.

Dividing by this value of $r$, we obtain for the other value,

$$r = 2R\cos v.$$

This value of $r$ will be positive when the $\cos v$ is positive; and negative, when the $\cos v$ is negative. But the negative values of the radius-vector must be rejected, since they cannot enter into the analysis.

The figure also indicates the same result. For, the $\cos v$ is positive in the first and fourth quadrants: hence, the radius-vector is positive when it falls in the first or fourth angle, The $\cos v$ is negative in the second and third quadrants: hence, the radius-vector is negative when it falls in the second or third angle.

Now, for $v = 0$, the $\cos v = 1$, and we have

$$r = 2R = B'B.$$

When $v$ increases from 0 to 90°, the radius-vector continues positive and determines all the points in the semi-circumference $B'CB$.

This may also be verified. For, in the right-angled riangle $B'CB$,

$$B'C = B'B\cos BB'C:$$

that is, $\quad r = 2R\cos v.$

When $v$ becomes equal to 90°, $\cos v = 0$, and $r$ becomes 0.

The radius-vector then becomes tangent to the circumference, since the two points in which it before cut it, have united.

From $v = 90°$ to $v = 270°$, the $\cos v$ is negative; and there is no point of the curve found in either the second or third angle.

From $v = 270°$ to $v = 360°$, the $\cos v$ is positive, and the radius-vector will determine all the points of the semicircumference below the axis of abscissas.

*Scholium* 2. If the pole be placed at the point $B$, whose co-ordinates are

$$a = +R, \qquad b = 0,$$

the equation will become

$$r = -2R \cos v.$$

In this equation the radius-vector will be negative when $\cos v$ is positive, and positive when the $\cos v$ is negative. Hence, the radius-vector will not give points of the curve from $v = 0$ to $v = 90°$.

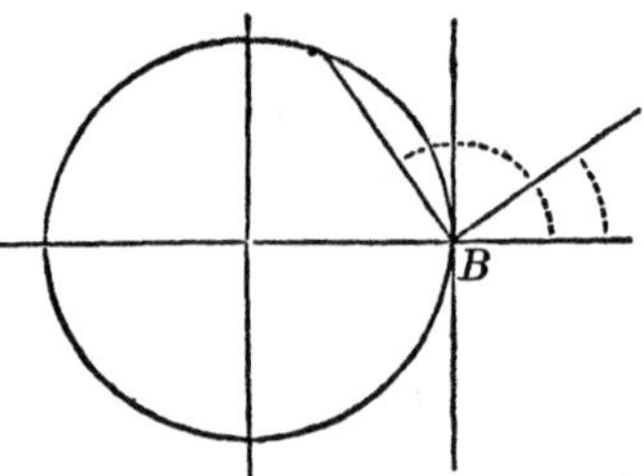


It will give points of the curve from $v = 90°$ to $v = 270°$ · and it will again fail to determine a curve from $v = 270°$ to $v = 360°$. The figure verifies these results.

*Scholium* 3. If we place the pole at the centre the equations for transformation will become,

$$x = r \cos v, \qquad y = r \sin v.$$

# BOOK IV.

## *Of the Ellipse.*

1. An *ellipse* is a curve in which the sum of two straight lines, drawn from any one of its points to two fixed points, is constantly equal to a given line.

Thus, if $F$ and $F'$ be the two fixed points, and $AB$ the given line; then, if $PF + PF'$ is constantly equal to $AB$ for every position of the point $P$, the curve $APBP$ will be an ellipse.

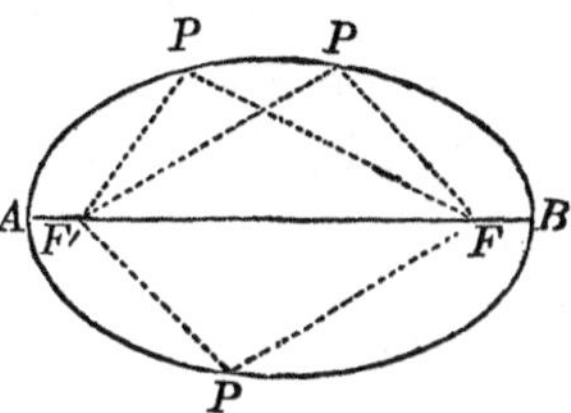


2. The fixed points $F$ and $F'$ are called *foci* of the ellipse.

3. The definition of an ellipse affords an easy method of describing it mechanically. Take a thread, longer than the distance $F'F$ and fasten its two extremities, the one at $F$, and the other at $F'$. Place a pencil against the thread, and move it around the points $F$, $F'$, keeping the thread constantly stretched, the point of the pencil will describe an ellipse.

### PROPOSITION I. PROBLEM.

*To find the equation of an ellipse.*

Let $F$ and $F'$ be the foci, and denote the distance between them by $2c$. Let $P$ be any point of the curve, and designate the distance $FP$ by $r$, and $F'P$ by $r'$; and let $2A$ represent the given line, to which the sum $FP + F'P$ is to be equal

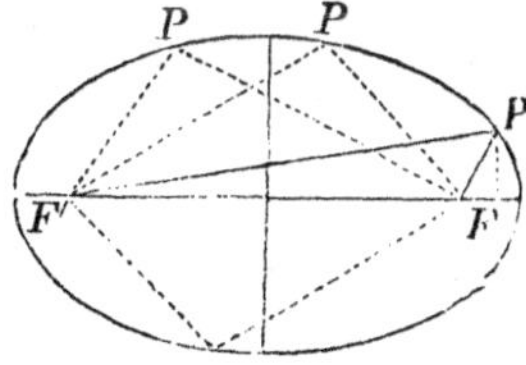

Through $C$, the middle point of $F'F$, draw $CD$ perpendicular to $F'F$, and let $C$ be the origin of a system of rectangular co-ordinates, of which $AB$, $DD'$ are the axes. Let $x$ and $y$ represent the co-ordinates of the point $P$.

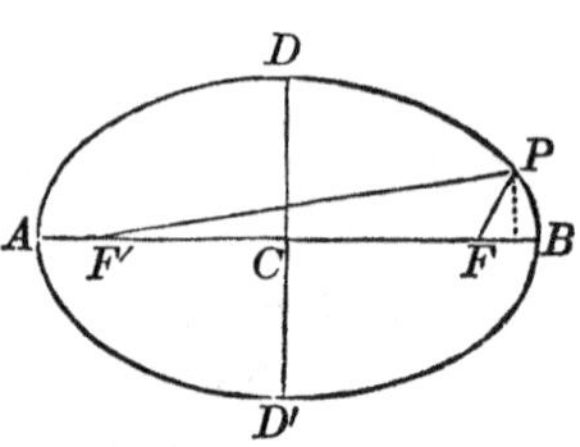


The square of the distance between any two points of which the co-ordinates are $x$, $y$, and $x'$, $y'$, is (Bk. II. Prop. III),

$$(y-y')^2+(x-x')^2.$$

If this line passes through the point $F$, of which the co-ordinates are $y'=0$, and $x'=c$, we shall have,

$$\overline{FP}^2=r^2=y^2+(x-c)^2;$$

and if we pass the line through the point $F'$, of which the co-ordinates are $y'=0$ and $x'=-c$, we shall have

$$r'^2=y^2+(x+c)^2.$$

If we add and subtract these two equations, we obtain

$$r^2+r'^2=2(y^2+x^2+c^2), \quad \text{and} \quad r'^2-r^2=4cx;$$

the last of which may be placed under the form,

$$(r'+r)(r'-r)=4cx.$$

But we have, from the property of the ellipse,

$$r+r'=2A.$$

Substituting for $r'+r$ its value $2A$, we have,

$$r'-r=\frac{2cx}{A}.$$

Combining this with the equation,

$$r' + r = 2A,$$

we obtain

$$r' = A + \frac{cx}{A}, \quad \text{and} \quad r = A - \frac{cx}{A}.$$

Squaring these values, and substituting in the equation of which the first member is $r^2 + r'^2$, and there results

$$A^2 + \frac{c^2x^2}{A^2} = y^2 + x^2 + c^2,$$

or $$A^2(y^2 + x^2) - c^2x^2 = A^2(A^2 - c^2),$$

which is the equation of the ellipse.

It is, however, most convenient to have the equation of the ellipse expressed in terms of the co-ordinates of its points, and the distances which the curve cuts off from the co-ordinate axes.

To place the equation under this form, let us make $x = 0$: this will give

$$y^2 = A^2 - c^2,$$

or $$y = \pm\sqrt{A^2 - c^2},$$

which is the value of $CD$ or $CD'$; and since $c$ is less than $A$ this value is always real.

If we represent $CD$ or $CD'$, by $\pm B$, we shall have

$$B^2 = A^2 - c^2,$$

or $$c^2 = A^2 - B^2.$$

Substituting this value of $c^2$ in the equation of the curve, it reduces to

$$A^2y^2 + B^2x^2 = A^2B^2,$$

in which, if we make $y=0$, we shall have

$$x = \pm A = CB \quad \text{or} \quad CA.$$

*Scholium* 1. The point $C$, which is equidistant from $F$ and $F'$, is called the *centre* of the ellipse.

Every straight line passing through the centre, and terminating in the curve, is called a *diameter*.

The diameter $AB$, which passes through the foci, is called the *transverse* axis. And since $2CA$, or $AB$, is equal to $2A$, it follows, that *the sum of the two lines drawn from any point of the curve to the foci, is equal to the transverse axis.*

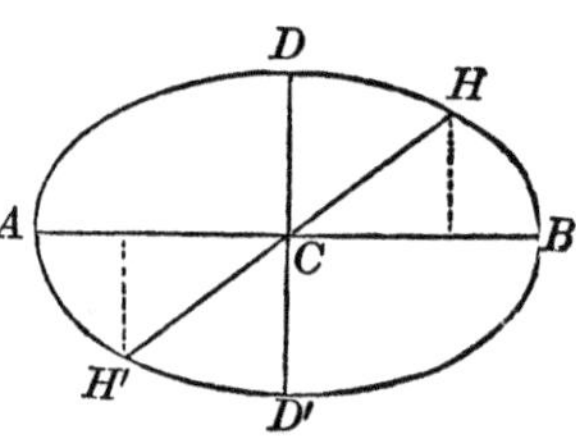


The diameter $DD'$, which is perpendicular to the transverse axis, is called the *conjugate axis.*

In the equation of the ellipse,

$$A^2y^2 + B^2x^2 = A^2B^2,$$

$A$ and $B$ represent the semi-axes, and $x$ and $y$ the general co-ordinates of the curve. It is called, *the equation of the ellipse referred to its centre and axes.*

*Scholium* 2. If through the centre of the ellipse any line be drawn, its equation will be of the form

$$y = ax.$$

If we combine this equation with the equation of the ellipse

$$A^2y^2 + B^2x^2 = A^2B^2,$$

we shall obtain the co-ordinates of the points $H$ and $H'$, in which the diameter intersects the curve.

If we designate the co-ordinates of $H$ by $x'$, $y'$, and the co-ordinates of $H'$ by $x''$, $y''$, we shall find, after eliminating

$$x' = AB\sqrt{\frac{1}{A^2a^2+B^2}}, \qquad y' = ABa\sqrt{\frac{1}{A^2a^2+B^2}},$$

$$x'' = -AB\sqrt{\frac{1}{A^2a^2+B^2}}, \qquad y'' = -ABa\sqrt{\frac{1}{A^2a^2+B^2}}.$$

But since the co-ordinates of the point $H'$ are the same as those of $H$, excepting that their signs are both negative it follows that $CH = CH'$: that is, *every diameter of an ellipse is bisected at the centre.*

*Scholium* 3. If $B$ be made equal to $A$, the equation of the ellipse will become,

$$y^2 + x^2 = A^2,$$

which is the equation of a circle: hence, *the ellipse becomes the circle when its axes become equal to each other.*

*Scholium* 4. The distance between the foci has been represented by $2c$: hence, the distance from the centre to either focus is equal to $c$. But we have seen that,

$$c^2 = A^2 - B^2,$$

hence, $$c = \pm\sqrt{A^2 - B^2}.$$

This distance divided by the semi-tranverse axis, is called the *eccentricity* of the ellipse; that is,

$$\frac{\sqrt{A^2 - B^2}}{A},$$

is the eccentricity.

When the ellipse becomes a circle the eccentricity is nothing.

*Scholium* 5. The equation of the ellipse,

$$A^2y^2 + B^2x^2 = A^2B^2,$$

may be put under the form

$$A^2y^2 + B^2x^2 - A^2B^2 = 0,$$

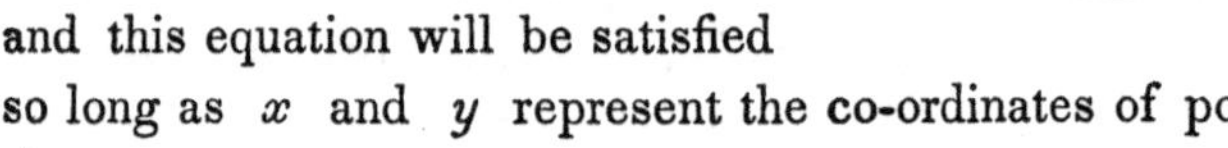

and this equation will be satisfied so long as $x$ and $y$ represent the co-ordinates of points of the curve.

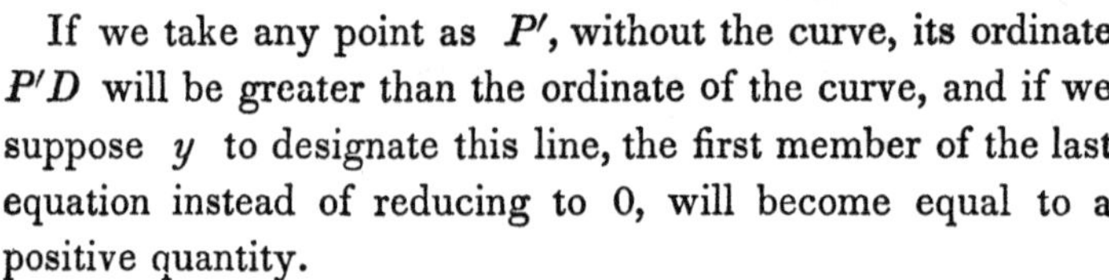

If we take any point as $P'$, without the curve, its ordinate $P'D$ will be greater than the ordinate of the curve, and if we suppose $y$ to designate this line, the first member of the last equation instead of reducing to 0, will become equal to a positive quantity.

If, on the contrary, we take a point $P''$ within the curve, its ordinate $P''D$ will be less than the ordinate of the curve, and if we designate this line by $y$, the first member of the last equation will become negative.

The following analytical conditions will, therefore, determine the position of a point with respect to the curve of the ellipse;

| | |
|---|---|
| without the ellipse, | $A^2y^2 + B^2x^2 - A^2B^2 > 0,$ |
| on the ellipse, | $A^2y^2 + B^2x^2 - A^2B^2 = 0,$ |
| within the ellipse, | $A^2y^2 + B^2x^2 - A^2B^2 < 0.$ |

*Scholium* 6. If we place the equation of the ellipse under the form

$$y = \pm\frac{B}{A}\sqrt{A^2 - x^2},$$

we see that every value of $x$, whether plus or minus, will give two equal values for $y$, with contrary signs: hence, the curve is symmetrical with respect to the transverse axis.

We also see, that if $x$ be made greater than $A$, whether it be taken plus or minus, the value of $y$ will become imaginary: hence, the curve will be limited both in the direction of $x$ positive and $x$ negative.

If we place the equation under the form

$$x = \pm \frac{A}{B}\sqrt{B^2 - y^2},$$

we see that for every value of $y$ there will be two equal values of $x$, with contrary signs: hence, the curve will be symmetrical with respect to the conjugate axis. It will also be limited in the direction of $y$ positive and $y$ negative.

*Scholium* 7. Any line as $HP$ or $HP'$ which represents the value of $y$ for any given abscissa $CH$, is called an *ordinate* of the ellipse, and is said to be an ordinate to the transverse axis.

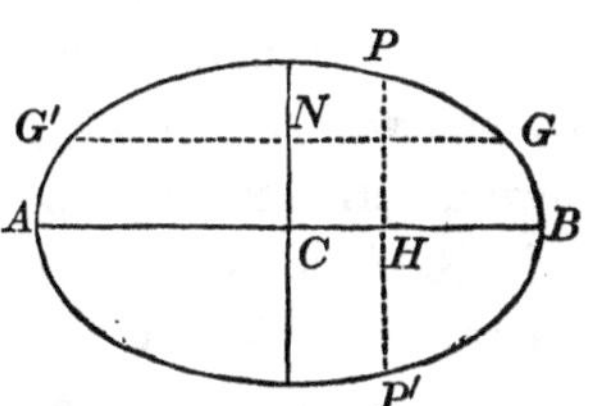


Every line, as $NG$ or $NG'$ which represents the value of $x$ for a given value of $y$, is also an ordinate of the ellipse, but it is an ordinate to the conjugate axis.

*Scholium* 8. If we transfer the origin of co-ordinates from the centre $C$, to $A$, one extremity of the transverse axis, the equations of transformation will reduce to (Bk. II, Prop. X)

$$x = -A + x', \qquad y = y'.$$

Substituting these values in the equation of the ellipse, it reduces to

$$A^2y'^2 + B^2x'^2 - 2B^2Ax' = 0$$

which may be put under the form

$$y'^2 = \frac{B^2}{A^2}(2Ax' - x'^2), \qquad \text{or,} \qquad y^2 = \frac{B^2}{A^2}(2Ax - x^2),$$

by omitting the accents; and this is the equation of the ellipse referred to the vertex of the transverse axis as an origin of co-ordinates.

*Scholium* 9. The property, that the sum of the two lines drawn from any point of the curve to the foci, is equal to the transverse axis, affords an easy method of describing the ellipse by points, when the transverse axis and foci are known.

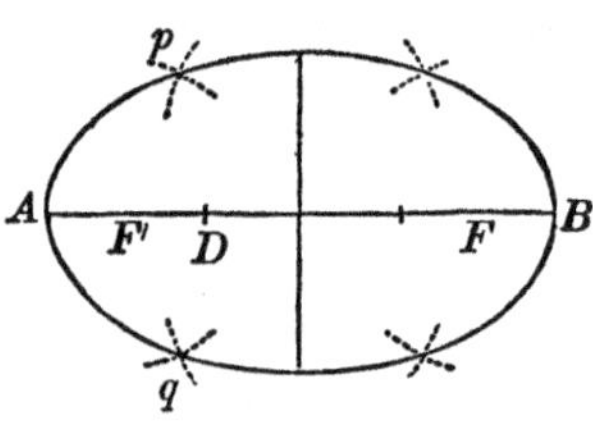


Let $AB$ be the transverse axis of an ellipse, and $F$ and $F'$ the foci. Take in the dividers any portion of the transverse axis, as $AD$, and with the focus $F'$ as a centre describe the arcs $p$ and $q$. With $BD$, the remaining part of the transverse axis, as a radius, and the other focus $F$ as a centre, describe two other arcs intersecting the former: the points of intersection will be points of the curve.

If with the radius $AD$, two arcs be described from the focus $F$, and with the radius $BD$ two arcs be described from the focus $F'$, these arcs will also determine, by their intersections, two points of the curve.

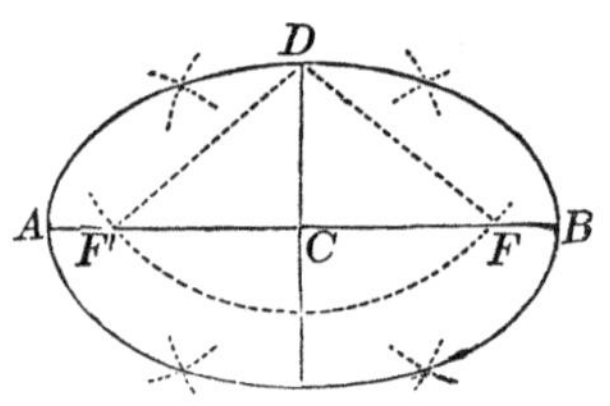


*Scholium* 10. If from either vertex of the conjugate axis, as $D$, the lines $DF'$, $DF$, be drawn to the foci, they will be equal to each other.

For, in the two right-angled triangles $F'CD$, $FCD$, $CF'$ is equal to $CF$, and $CD$ is common: hence, the hypotheneuse $F'D$ is equal to $FD$ (Geom. Bk. I, Prop. V).

But $F'D + DF$ is equal to $AB$: hence, $DF'$ or $DF$ is equal to $CB$. If, therefore, with either vertex of the conjugate axis as a centre, and with a radius equal to half the transverse axis, the circumference of a circle be described, it will intersect the transverse axis at the foci.

PROPOSITION II. THEOREM.

*The squares of the ordinates to either axis of the ellipse are to each other as the rectangles of the corresponding segments into which they divide the axis.*

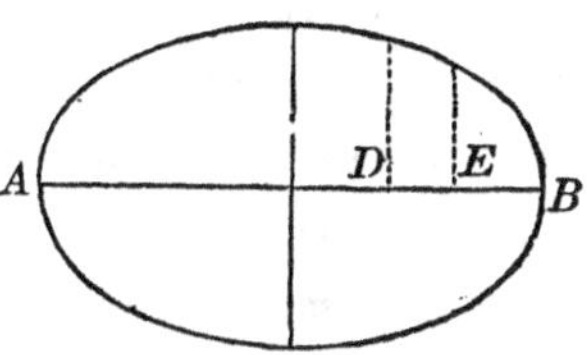


The equation of the ellipse referred to the vertex $A$, as the origin of co-ordinates (Prop. I, Sch. 8) is,

$$y^2 = \frac{B^2}{A^2}(2A - x)x.$$

If we designate a particular ordinate by $y'$, and its abscissa by $x'$, and a second ordinate by $y''$, and its abscissa by $x''$, we shall have,

$$y'^2 = \frac{B^2}{A^2}(2A - x')x', \quad \text{and} \quad y''^2 = \frac{B^2}{A^2}(2A - x'')x''.$$

Dividing one equation by the other, we obtain

$$\frac{y'^2}{y''^2} = \frac{(2A - x')x'}{(2A - x'')x''},$$

or $$y'^2 : y''^2 :: (2A - x')x' : (2A - x'')x''.$$

But $2A$ represents the transverse axis $AB$, and since $x' = AD$, $2A - x' = DB$; therefore, $(2A - x')x'$ represent the rectangle of the segments $AD$, $DB$. In like manner it may be shown, that $(2A - x'')x''$ is equal to the product of the segments $AE$, $EB$.

It may be proved in a similar manner, that the squares of the ordinates to the conjugate axis, are to each other as the rectangles of the segments.

PROPOSITION III. THEOREM.

*If on the transverse axis of an ellipse the circumference of a circle be described, and if on the conjugate axis the circumference of a circle be also described: then,*

1st. *Any ordinate of the ellipse, drawn to the transverse axis, will be to the corresponding ordinate of the circle, as the semi-conjugate axis to the semi-transverse axis;* and

2dly. *Any ordinate drawn to the conjugate axis, will be to the corresponding ordinate of the circle, as the semi-transverse axis to the semi-conjugate axis.*

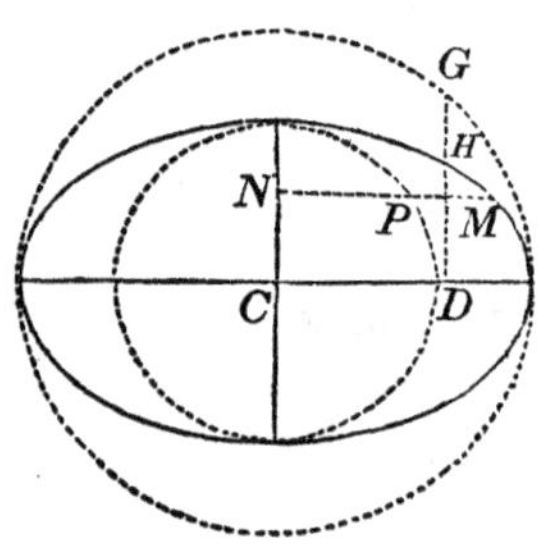


Let $y'$ and $Y'$ designate the ordinates $DH$, $DG$, corresponding to the same abscissa $CD$, which we will designate by $x'$. We shall then have,

$$y'^2 = \frac{B^2}{A^2}(A^2 - x'^2),$$

and $Y'^2 = A^2 - x'^2$;

hence, $$\frac{Y'^2}{y'^2} = \frac{A^2}{B^2}, \quad \text{or} \quad \frac{Y'}{y'} = \frac{A}{B}.$$

Therefore, $y' : Y' :: B : A.$

To prove the second part of the proposition, denote the ordinates $NM$, $NP$, corresponding to the same point $N$ of the conjugate axis, by $x'$ and $X'$, and designate $CN$ by $y'$: we shall then have,

$$x'^2 = \frac{A^2}{B^2}(B^2 - y'^2),$$

and $$X'^2 = B^2 - y'^2;$$

hence, $$\frac{X'^2}{x'^2} = \frac{B^2}{A^2}, \quad \text{or} \quad \frac{X'}{x'} = \frac{B}{A}$$

Therefore. $x' : X' :: A : B$

*Corollary.* From what has been demonstrated, it is plain that every point of the ellipse is within the circumference described on the transverse axis, and without the circumference described on the conjugate axis. Hence, *the transverse axis is the greatest diameter of the ellipse, and the conjugate axis is the least.*

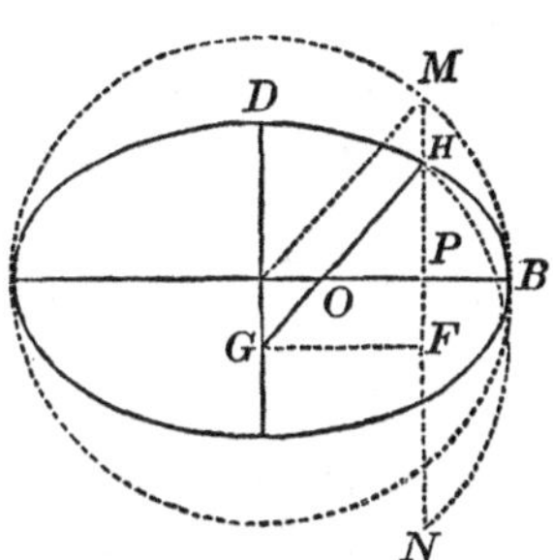


*Scholium.* The last proposition affords the following easy method of describing an ellipse when the axes are known.

Let $GH$ be a ruler, or slip of paper, equal in length to $A$, the semi-transverse axis. Mark on it a distance $HO$ equal to $B$; then will $GO$ be equal to $A - B$, or the difference of the semi-axes.

Place the ruler in such a manner, that the extremity $G$ shall fall on the conjugate axis, and the point $O$, on the transverse axis, and move it around, keeping these points constantly on the axes; the point $H$ will describe the arc $BHD$ of the ellipse. By placing the ruler in the other angles, the entire curve may be described.

To prove that $H$ is a point of the ellipse, describe on the transverse axis a semi-circumference, and with $G$ as a centre, and $GH = A$ as a radius, describe the arc $HN$. Now $MP$ will be equal to $HF$. But by similar triangles, we have

$$GH : OH :: FH \quad PH,$$

that is $$A : B :: PM : PH;$$

hence, the point $H$ is on the ellipse.

PROPOSITION IV. THEOREM.

*If through the vertices of the transverse axis two supplementary chords be drawn, the product of the tangents of the angles which they form with it, on the same side, will be negative and equal to the square of the ratio of the semi-axes.*

The equation of a straight line passing through the point $A$, of which the co-ordinates are

$$x' = -A, \quad y' = 0, \quad \text{is}$$

$$y = a'(x + A);$$

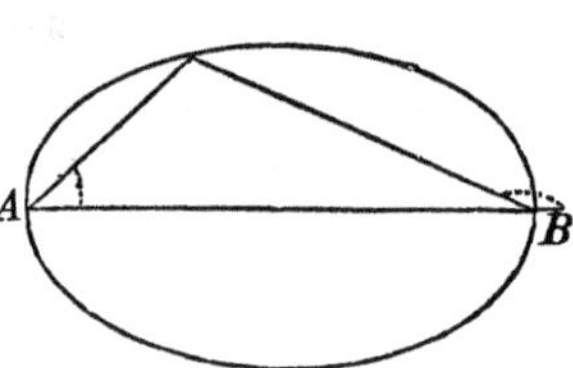


and of a line passing through $B$, of which the co-ordinates are $x' = +A$, $y' = 0$,

$$y = a(x - A).$$

If these lines intersect each other, we have

$$y^2 = aa'(x^2 - A^2);$$

but if they intersect on the curve of the ellipse, $x$ and $y$ must satisfy the equation

$$y^2 = \frac{B^2}{A^2}(A^2 - x^2) = -\frac{B^2}{A^2}(x^2 - A^2).$$

By combining these equations, we find

$$aa' = -\frac{B^2}{A^2}.$$

*Scholium* 1. In the equation

$$aa' = -\frac{B^2}{A^2},$$

there are two undetermined quantities, $a$ and $a'$: hence, an infinite number of supplementary chords may be drawn through the extremities of the diameter $AB$

If, however, a value be assigned to $a$, or $a'$, that is, if one of the supplementary chords be given in position, the equation of condition will determine the other, and therefore the corresponding supplementary chord may also be drawn.

*Scholium* 2. If the ellipse becomes a circle, we shall have

$$aa' = -1,$$

or

$$aa' + 1 = 0;$$

which shows, that the supplementary chords are perpendicular to each other, a property before proved (Bk. III Prop II).

*Scholium* 3. The supplementary chords which are drawn through the extremities of the transverse axis form with each other an obtuse angle.

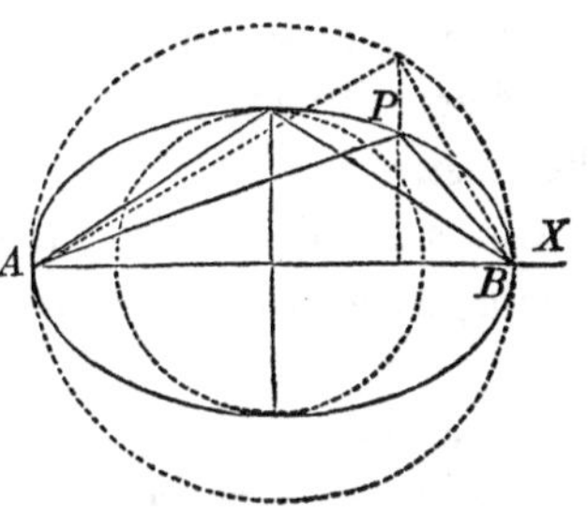


For, if on the transverse axis the circumference of a circle be described, all the points of the ellipse will be within it.

Hence, the angle formed by two supplementary chords of the ellipse will be within the angle formed by the corresponding supplementary chords of the circle, and will therefore be the greater angle. But the angle formed in the circle is a right angle; hence, the angle formed by the supplementary chords in the ellipse is obtuse.

It may be shown, in a similar manner, that the angle formed by the supplementary chords drawn through the extremities of the conjugate axis, is acute.

To determine at what point of the ellipse the maximum angle is formed, denote the angle $PBX$ by $\alpha'$, the angle $PAB$ by $\alpha$, and their tangents by $a'$ and $a$: designate the angle $APB$ by $V$. Then, $V = \alpha' - \alpha$, and

$$\text{tang } V = \frac{a' - a}{1 + aa'}, \qquad \text{(Trig. Art. XXV).}$$

But since $a'$ is negative, we may give to it its sign, and place the numerator under the form $-(a' + a)$; and since

$$aa' = -\frac{B^2}{A^2},$$

we have

$$\text{tang } V = \frac{-(a' + a)}{1 - \frac{B^2}{A^2}}.$$

Now, as $V$ exceeds 90° it will have the greatest value when its tangent has the least value. But since the denominator in the second member is a constant quantity, that member will be the least when the numerator $-(a' + a)$ is the least.

But, since the product $a'a$ has a given value, the sum will be the least when $a' = a$, or by replacing the minus sign, when $-a' = a$.

Hence, when the angles which the supplementary chords form with the transverse axis, on the same side, are supplements of each other, the angle formed by the chords will be the greatest possible, and their point of intersection will be at one of the vertices of the conjugate axis.

It might be proved, in a similar manner, that if the supplementary chords were drawn through the vertices of the conjugate axis, the least angle would be formed when their point of intersection is found at one of the vertices of the transverse axis.

## PROPOSITION V. PROBLEM.

*To find the equation of a straight line which shall be tangent to an ellipse.*

The equation of the ellipse is

$$A^2y^2 + B^2x^2 = A^2B^2.$$

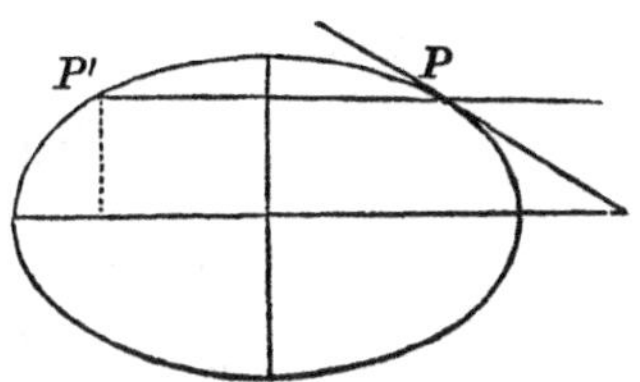


Take any point of the curve, as $P$, and designate its co-ordinates by $x''$, $y''$. Through this point draw a secant line: its equation will be of the form

$$y - y'' = a(x - x''):$$

it is now required to find the value of $a$ when the secant line $PP'$ becomes tangent to the ellipse.

Since the point $P$ is on the curve, we shall have,

$$A^2y''^2 + B^2x''^2 = A^2B^2.$$

Subtracting this from the equation of the ellipse, and we have,

$$A^2(y^2 - y''^2) + B^2(x^2 - x''^2) = 0,$$

or $$A^2(y + y'')(y - y'') + B^2(x + x'')(x - x'') = 0;$$

in this equation $x$ and $y$ are the co-ordinates of any point of the ellipse.

If this equation be combined with the equation of the secant, $x$ and $y$ in the resulting equation, will be the co-ordinates of $P'$, the second point in which the secant intersects the ellipse. They are most readily combined by substituting for $y - y''$ its value taken from the equation of the secant. Substituting, we obtain,

$$A^2(y+y'')a(x-x'')+B^2(x+x'')(x-x'')=0;$$

and dividing by $x-x''$, we have

$$A^2(y+y'')a+B^2(x+x'')=0.$$

If now we suppose $P'$ to move towards $P$ we shall have at the time it coincides with it,

$$x=x'', \quad \text{and} \quad y=y'';$$

and the last equation will give

$$a=-\frac{B^2x''}{A^2y''}.$$

Substituting this value in the equation of the line passing through $P$, and we have

$$y-y''=-\frac{B^2x''}{A^2y''}(x-x''),$$

or by reducing

$$A^2yy''-A^2y''^2=-B^2xx''+B^2x''^2$$

or $$A^2yy''+B^2xx''=A^2y''^2+B^2x''^2,$$

or, $$A^2yy''+B^2xx''=A^2B^2,$$

in which $y$ and $x$ are the general co-ordinates of the tangent line.

*Scholium* 1. We can easily prove, as in the circle, that every point of this line, except the point of contact, lies without the ellipse.

We have the equations

$$A^2yy''+B^2xx''=A^2B^2,$$

$$A^2y''^2+B^2x''^2=A^2B^2.$$

If we subtract twice the first equation from the second, and then add $A^2y^2+B^2x^2$ to each member of the resulting equation, it may be put under the form

$$A^2(y-y'')^2+B^2(x-x'')^2=A^2y^2+B^2x^2-A^2B^2.$$

The first member of this equation is positive; therefore the second member is positive: hence, every point of the tangent line lies without the ellipse, except the point of contact, the co-ordinates of which reduce both members to 0 (Prop. I, Sch. 5)

*Scholium* 2. To find the point in which the tangent intersects the axis of abscissas, we make $y=0$ in the equation

$$A^2yy''+B^2xx''=A^2B^2,$$

which gives $$x=\frac{A^2}{x''};$$

and this value of $x$ is equal to $CT$.

If from $CT$ we subtract $CR$, which is designated by $x''$, we shall obtain

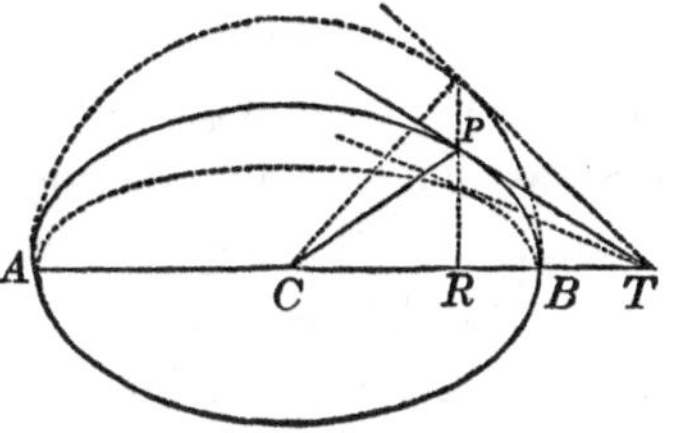


$$TR=\frac{A^2}{x''}-x''=\frac{A^2-x''^2}{x''}.$$

This distance is called the *sub-tangent*, and is, as its name implies, the projection of the tangent on the axis of abscissas. This expression for the sub-tangent is independent of the conjugate axis, and will therefore be the same, for all the ellipses having the same transverse axis $AB$, and the points of tangency in the same perpendicular $RP$.

If we determine, in like manner, the sub-tangent on the conjugate axis, it will be found to be independent of the transverse axis.

*Scholium* 3. The property just proved, offers an easy construction for drawing a tangent line to an ellipse at a given point.

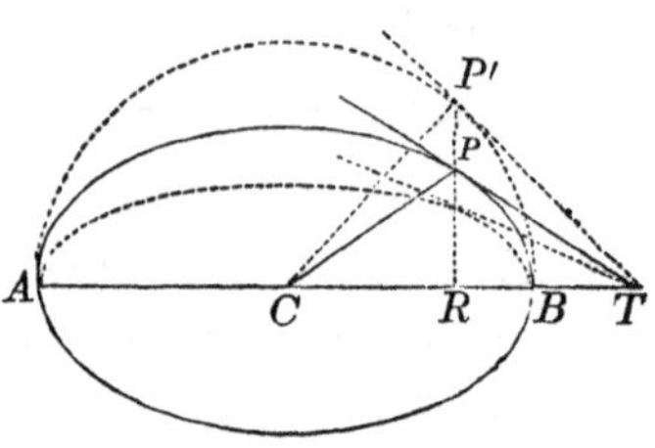


Let $P$ be the given point. On $AB$, describe a semi-circumference, and through $P$, draw $PR$ perpendicular to $AB$, and produce it till it meets the circumference at $P'$. Through $P'$, draw a tangent line to the circumference of the circle, and from $T$, where it meets $AB$ produced, draw $TP$, and it will be tangent to the ellipse at $P$.

The angle $CP'T$ being a right angle, the angle $CPT$, which lies within it, is obtuse. Hence, the angle formed by a tangent line, and the diameter passing through the point of contact, will be obtuse if estimated on one side of the diameter, and acute if estimated on the other.

If, however, the point of contact is at either vertex of the conjugate axis, the tangent line to the ellipse will become parallel to the tangent line to the circle, and consequently, perpendicular to the conjugate axis. Or, if the point of tangency be at either vertex of the transverse axis, the tangent line to the ellipse will coincide with the tangent line to the circle, and will therefore be perpendicular to the transverse axis.

### PROPOSITION VI. PROBLEM.

*To find the equation of a normal line to the ellipse.*

Since the normal passes through the point of tangency, its equation will be of the form

$$y - y'' = a'(x - x'');$$

and since it is perpendicular to the tangent, we shall have

$$aa' + 1 = 0.$$

But we have already found

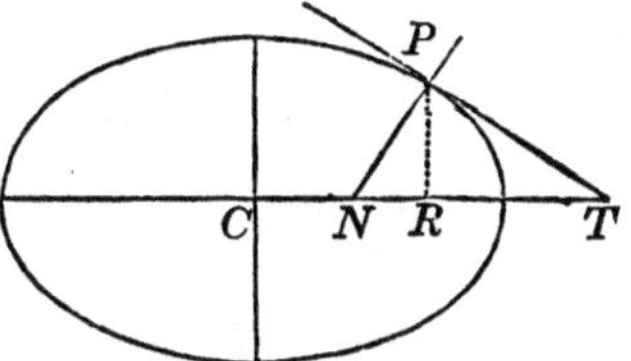


$$a = -\frac{B^2x''}{A^2y''};$$

hence, $a' = \frac{A^2}{B^2}\frac{y''}{x''}$.

Substituting this value, and the equation of the normal will become

$$y - y'' = \frac{A^2}{B^2}\frac{y''}{x''}(x - x'').$$

*Scholium.* To find the point in which the normal intersects the axis of $X$, make $y = 0$, and we have

$$CN = x = \frac{A^2 - B^2}{A^2}x''.$$

If we subtract this value from $CR$, which is designated by $x''$, we shall find

$$NR = \frac{B^2x''}{A^2}$$

which is called the *sub-normal.*

### PROPOSITION VII. THEOREM.

*If one of the supplementary chords of an ellipse be parallel to a tangent line to the curve, the other will be parallel to the diameter which passes through the point of contact:* and conversely,

*If one of the chords be parallel to the diameter which passes through the point of contact, the other will be parallel to the tangent line*

The equation of a line passing through the centre of the ellipse is of the form

$$y = a'x.$$

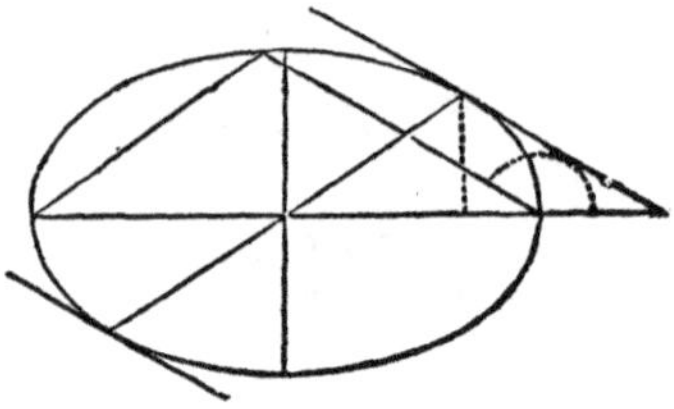

The condition of its passing through the point of contact, will give

$$y'' = a'x'';$$

hence $$a' = \frac{y''}{x''}.$$

But we have found the tangent of the angle which the tangent line makes with the transverse axis, to be

$$a = -\frac{B^2x''}{A^2y''}.$$

Multiplying the members of these equations together, we obtain

$$aa' = -\frac{B^2}{A^2}.$$

By comparing this equation with the equation in Prop. IV, we see, that the product of the tangents of the angles which the diameter and tangent line make with the transverse axis, is equal to the product of the tangents of the angles which the supplementary chords form with the axis Hence if in these equations we make

$$a = a,$$

we shall have $$a' = a';$$

that is, if one chord is parallel to the tangent, the other will be parallel to the diameter passing through the point of contact.

Or, if we make

$$a' = a'$$

we shall have $$a = a;$$

that is, if one of the chords be made parallel to the diameter the other will be parallel to the tangent.

*Scholium* 1. The equation of the ellipse being symmetrical with respect to its axes, the properties which have been demonstrated with respect to the tangent, diameter, and the supplementary chords, when referred to the transverse axis, are generally true with respect to the conjugate axis.

*Scholium* 2. These properties also afford an easy method of drawing a tangent line to an ellipse at a given point of the curve.

Let $C$ be the centre of an ellipse, $AB$ the transverse axis, and $P$ the point of the curve at which the tangent is to be drawn.

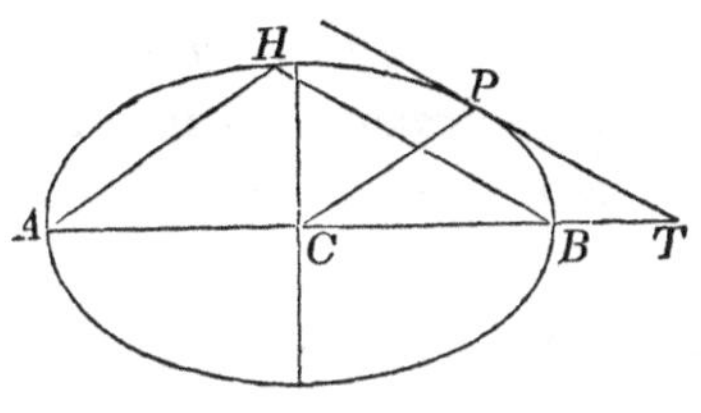


Through $P$ draw the semi-diameter $PC$, and through $A$ draw the supplementary chord $AH$ parallel to it. Then draw the other supplementary chord $BH$, and through $P$ draw $PT$ parallel to $BH$; then will $PT$ be the tangent required.

*Scholium* 3. The same property will likewise enable us to draw a tangent line to an ellipse which shall be parallel to a given line.

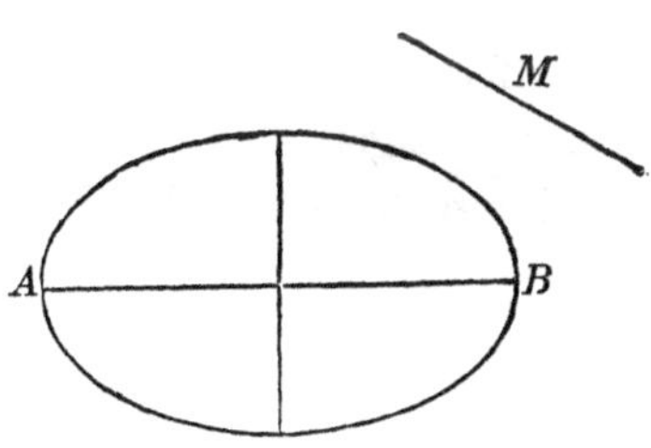


Let $AB$ be the transverse axis of the ellipse, and $M$ the given line.

Through the vertex $B$ draw the supplementary chord $BG$ parallel to $M$.

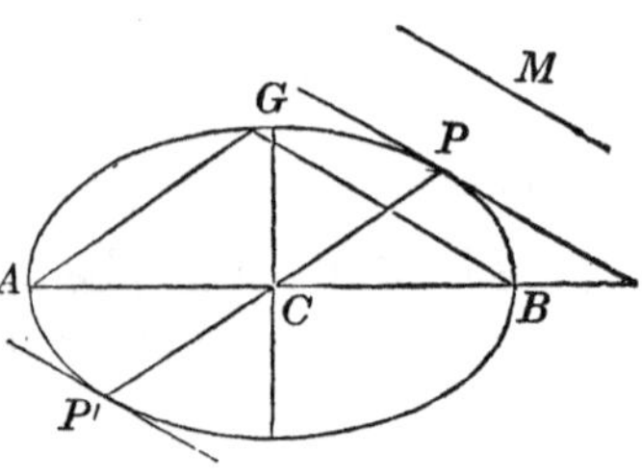


Then draw $AG$, and through the centre $C$ draw $CP$ parallel to $AG$ and produce it till it meets the ellipse again at $P'$. Through $P$ or $P'$ draw a parallel to $GB$, and it will be the tangent required.

We see, from this construction, that if two tangents be drawn to the ellipse through the two extremities of the same diameter, they will be parallel to each other.

### PROPOSITION VIII. THEOREM.

*If a line be drawn tangent to an ellipse at any point, and two lines be drawn from the same point to the two foci; then, the lines drawn to the foci will make equal angles with the tangent.*

Let $C$ be the centre of the ellipse, $PT$ the tangent line, and $PF$, $PF'$, the two lines drawn to the foci.

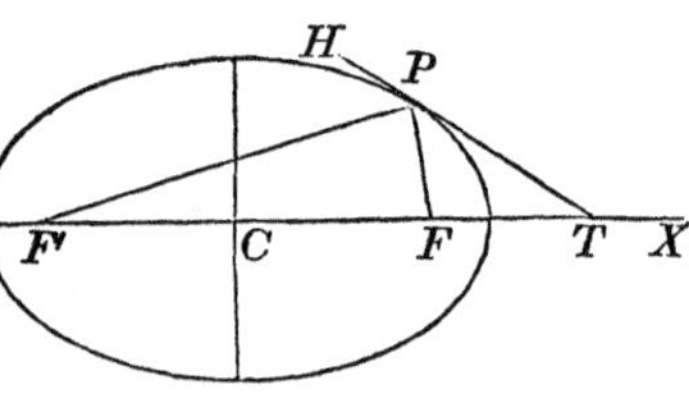


Denote the distance $CF=\sqrt{A^2-B^2}$ by $c$, $CF'$ by $-c$, the angle $FPT$ by $V$, and the tangents of the angles $PTX$, $PFT$, by $a$ and $a'$. We shall then have (Bk. II, Prop. VII),

$$\text{tang } V=\frac{a-a'}{1+aa'}.$$

But the equation of the line $FP$ passing through two

points of which the co-ordinates are $x'$, $y'$, $x''$, $y''$, is (Bk. II Prop. V),

$$y - y'' = \frac{y' - y''}{x' - x''}(x - x'').$$

But the co-ordinates of the focus $F$ are $x' = c$ $y' = 0$; hence we have,

$$y - y'' = \frac{-y''}{c - x''}(x - x''),$$

for the equation of $FP$; hence we obtain,

$$a' = \frac{-y''}{c - x''}.$$

But we also have

$$a = -\frac{B^2 x''}{A^2 y''}.$$

Substituting these values, and we obtain,

$$\text{tang } FPT = \frac{A^2 y''^2 + B^2 x''^2 - B^2 c x''}{A^2 c y'' - (A^2 - B^2) x'' y''},$$

which reduces to

$$\frac{B^2}{c y''}$$

by observing that the point of tangency is on the ellipse, and that $c^2 = A^2 - B^2$:

The equation of the line $PF$ will become the equation of the line $PF'$, by simply changing $+c$ into $-c$, for then the line would pass through the focus $F'$.

Hence, the angle $FPT$ will become the angle $F'PT$ by changing $+c$ into $-c$.

This gives the

$$\text{tang } F'PT = -\frac{B^2}{c y''}.$$

Now, since the angles $FPT$, $F'PT$, have equal tangents with contrary signs, they will be supplements of each other (Trig. Art. XIII).

Hence $FPT + F'PT = 180°$.

But, $F'PH + F'PT = 180°$;

therefore, $FPT = F'PH$.

*Corollary.* The normal being perpendicular to the tangent will bisect the angle included between the two lines drawn to the foci.

*Scholium* 1. The last proposition furnishes a simple construction for drawing a tangent line to an ellipse at a given point of the curve.

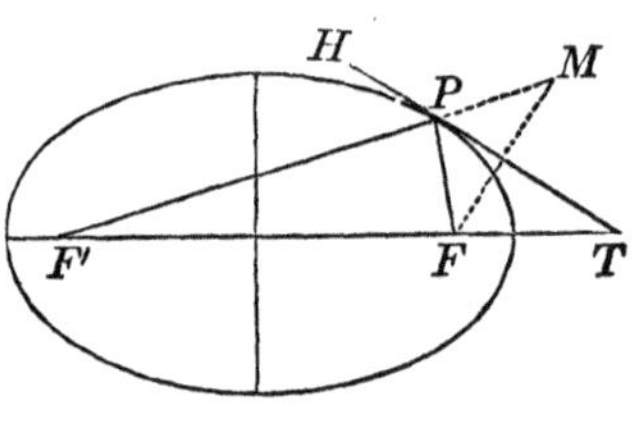


Let $P$ be the given point. From $P$ draw the lines $PF'$ and $PF$ to the foci. Produce $F'P$ until $PM$ shall be equal to $PF$, and draw $FM$. Then draw $PT$ perpendicular to $FM$ and it will be the tangent required, since it makes equal angles with the lines $PF'$ and $PF$.

*Scholium* 2. The same properties will also enable us to draw a tangent line to the ellipse, through a given point without the curve.

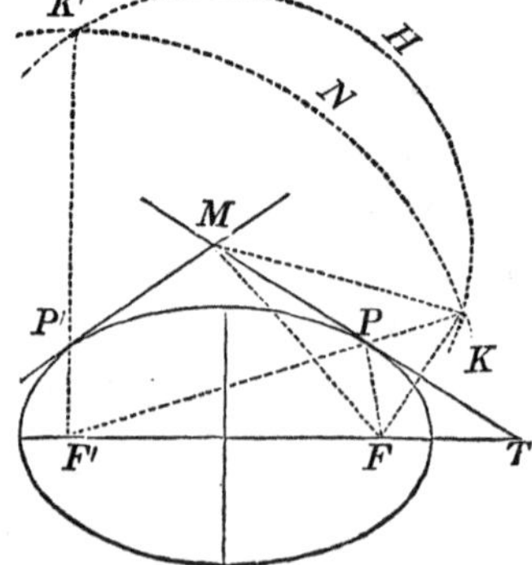


Let $M$ be the given point. With either focus, as $F'$, as a centre, and a radius equal to the transverse axis, describe the arc $KNK'$. Then, with $M$ as a centre, and a radius equal to $MF$, the distance to the other focus, describe the arc $FKHK'$, intersecting the former in $K$ and $K'$. Through $K$ draw $KF'$; and through

$P$, where it intersects the ellipse, draw the straight line $MP$, and it will be tangent to the ellipse at $P$.

For, since $P$ is a point of the ellipse $F'P + PF$ is equal to the transverse axis. But $F'P + PK$ is equal to the transverse axis, by construction. Hence, $PF = PK$.

Further, since the arc $FK$ is described from the centre $M$, $MF = MK$: hence, the line $MP$ has two of its points equally distant from the points $F$ and $K$: it is therefore perpendicular to $FK$ (Geom. Bk. I, Prop. XVI, Cor.); and since the triangle $FPK$ is isosceles, $MP$ will bisect the vertical angle $P$. The opposite angle $F'PM$, being equal $TPK$, is equal to $FPT$: hence, $MT$ is tangent to the ellipse.

*Scholium* 3. The two arcs $KHK'$, $KNK'$, will in general intersect each other in two points, $K$ and $K'$. There will, therefore, be two lines, $KF'$, $K'F'$, drawn to the focus $F'$, and two points of contact, $P$, $P'$. and consequently two tangent lines $MP$, $MP'$.

It may also be shown analytically, that two tangent lines can be drawn to the ellipse from a given point without the curve.

The equation of the tangent is

$$A^2yy'' + B^2xx'' = A^2B^2;$$

and if the tangent be made to pass through a given point, whose co-ordinates are $x'$, $y'$, the equation will become

$$A^2y'y'' + B^2x'x'' = A^2B^2.$$

But since the point of contact is on the curve, we also have

$$A^2y''^2 + B^2x''^2 = A^2B^2.$$

In these two equations all the quantities are known, except $x''$, $y''$, which may therefore be found, and since the equation of the tangent line is of the first degree, with respect to

$x''$, $y''$, the equation which results from the combination of the two equations will be of the second degree, and will therefore give two values for $x''$ and two values for $y''$, which values will be real, if the given point lies without the curve.

## *Of the Ellipse referred to its Conjugate Diameters.*

Two diameters of an ellipse are said to be conjugate, when either of them is parallel to the two tangent lines which may be drawn through the vertices of the other.

Since two supplementary chords may be drawn respectively parallel to a diameter and the tangent lines through its vertices (Prop. VII), it follows, that two supplementary chords may always be drawn parallel to any two conjugate diameters.

Therefore, if we designate by $a$ and $a'$ the tangents of the angles which two conjugate diameters make, respectively, with the transverse axis, these tangents must satisfy the equation,

$$aa' = -\frac{B^2}{A^2}.$$

Let us designate the corresponding angles by $\alpha$ and $\alpha$ We shall then have,

$$a = \frac{\sin \alpha}{\cos \alpha} \quad \text{and} \quad a' = \frac{\sin \alpha'}{\cos \alpha'}.$$

Substituting these values in the last equation, and reducing, we obtain

$$A^2 \sin \alpha \sin \alpha' + B^2 \cos \alpha \cos \alpha' = 0.$$

which expresses the relation between the angles which two conjugate diameters form with the transverse axis, and is called, the equation of condition for conjugate diameters.

In the equation of condition, $\alpha$ and $\alpha$ are undetermined. Hence, any value may be assigned, at pleasure, to either of them; and when assigned to one, the value of the other can be determined from the equation of condition.

If $\alpha = 0$, we shall have $\sin \alpha = 0$, and $\cos \alpha = 1$. Hence $B^2 \cos \alpha' = 0$, and consequently, $\cos \alpha' = 0$, or $\alpha' = 90°$ Therefore, when one of the conjugate diameters coincides with the transverse axis, the other will coincide with the conjugate axis. The axes, therefore, fulfil the condition of conjugate diameters, as they should do, since each is parallel to the tangents drawn through the vertices of the other.

The axes are the only conjugate diameters which are at right angles to each other.

For, if there are others,

$$\alpha' - \alpha = 90°, \quad \text{or} \quad \alpha' = 90° + \alpha,$$

which gives,

$$\sin \alpha' = \sin 90° \cos \alpha + \cos 90° \sin \alpha = + \cos \alpha,$$

$$\cos \alpha' = \cos 90° \cos \alpha - \sin 90° \sin \alpha = - \sin \alpha.$$

Substituting these values in the equation of condition,

$$A^2 \sin \alpha \sin \alpha' + B^2 \cos \alpha \cos \alpha' = 0$$

and we have,

$$(A^2 - B^2) \sin \alpha \cos \alpha = 0,$$

an equation which cannot be satisfied unless

$$\alpha = 0 \quad \text{or} \quad \alpha = 90.$$

If, however, we make $A = B$, the ellipse will become a circle, and then the equation of condition will be satisfied for all values of $\alpha$; which proves, that, *in the circle all conjugate diameters are at right angles to each other*

PROPOSITION IX. PROBLEM.

*To find the equation of the ellipse referred to its centre and conjugate diameters.*

The equation of the ellipse referred to its centre and axes is,

$$A^2y^2 + B^2x^2 = A^2B^2.$$

The formulas for passing from rectangular to oblique co-ordinates, the origin remaining the same, are (Bk. II, Prop. XI),

$$x = x' \cos\alpha + y' \cos\alpha', \qquad y = x' \sin\alpha + y' \sin\alpha'.$$

Squaring these values of $x$ and $y$, and substituting in the equation of the ellipse, we have

$$\left\{\begin{array}{c}(A^2\sin^2\alpha' + B^2\cos^2\alpha')y'^2 + (A^2\sin^2\alpha + B^2\cos^2\alpha)x'^2 \\ + 2(A^2\sin\alpha\sin\alpha' + B^2\cos\alpha\cos\alpha')x'y'\end{array}\right\} = A^2B^2.$$

But the condition that the new axes shall be conjugate diameters, gives

$$A^2 \sin\alpha \sin\alpha' + B^2 \cos\alpha \cos\alpha' = 0;$$

hence, the equation reduces to

$$(A^2\sin^2\alpha' + B^2\cos^2\alpha')y'^2 + (A^2\sin^2\alpha + B^2\cos^2\alpha)x'^2 = A^2B^2$$

But the equation can be placed under a more simple form by introducing the semi-conjugate diameters $CB'$ and $CD'$, which we will represent by $A'$ and $B'$.

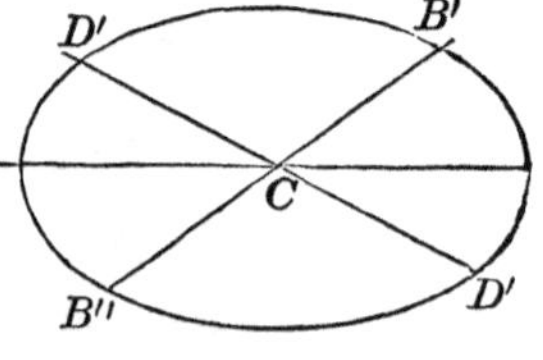


If we make $y' = 0$, we shall have

$$x'^2 = \frac{A^2B^2}{A^2\sin^2\alpha + B^2\cos^2\alpha} = CB'^2 = CB''^2 = A'^2.$$

If we make $x' = 0$, we shall have

$$y'^2 = \frac{A^2B^2}{A^2 \sin^2 \alpha' + B^2 \cos^2 \alpha'} = CD'^2 = CD''^2 = B'^2.$$

The denominators in the two last equations are the coefficients of $x'^2, y'^2$, in the equation from which they were deduced. Finding their values, and substituting them in that equation; and reducing, we obtain

$$A'^2 y'^2 + B'^2 x'^2 = A'^2 B'^2,$$

or omitting the accents of $x'$ and $y'$, since they are general variables, we obtain

$$A'^2 y^2 + B'^2 x^2 = A'^2 B'^2.$$

for the equation of the ellipse referred to its centre and conjugate diameters.

*Scholium* 1. The equation of the ellipse referred to its centre and conjugate diameters being of the same form as when referred to its centre and axes, it follows, that every value of $x$ will give two equal values of $y$ with contrary signs; and every value of $y$ will give two equal values of $x$ also with contrary signs: hence the ellipse is symmetrical with respect to either of its conjugate diameters: that is, *either diameter will bisect all chords drawn parallel to the other and terminated by the curve.*

*Scholium* 2. The last property points out the following method of finding the centre and axes of an ellipse when the curve is traced on a plane.

Draw two parallel chords and through the middle point of each draw a straight line, and it will be a diameter of the ellipse. Find a second diameter in the same manner and their point of intersection will be the centre $C$.

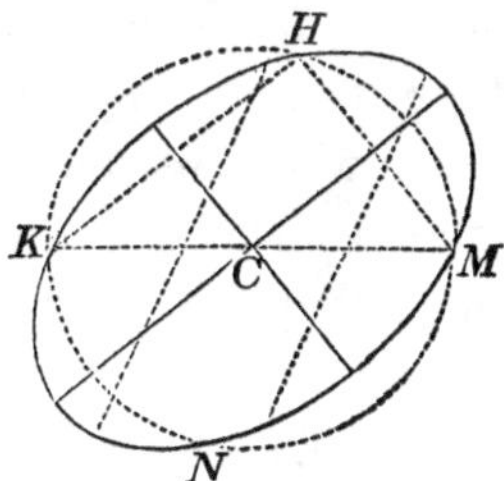

With $C$ as a centre, describe the circumference of a circle intersecting the ellipse at $H$, $K$, $N$ and $M$, and draw $HK$, and $HM$.

The lines drawn from $C$ perpendicular to $KH$, $HM$ will be the axes of the ellipse.

The reasons for this construction will readily suggest themselves to the reader.

*Scholium* 3. The equation of the ellipse referred to its conjugate diameters is,

$$A'^2y^2 + B'^2x^2 = A'^2B'^2.$$

If we designate any two ordinates by $y'$, $y''$, and the corresponding abscissas by $x'$, $x''$, we shall have,

$$\frac{y'^2}{y''^2} = \frac{(A' + x')(A' - x')}{(A' + x'')(A' - x'')},$$

$$y'^2 : y''^2 :: (A' + x')(A' - x') : (A' + x'')(A' - x'').$$

If the ordinates be drawn parallel to $AB$, it may be readily shown, that,

$$x'^2 : x''^2 :: (B' + y')(B' - y') : (B' + y'')(B' - y'').$$

Hence, *the squares of the ordinates to either one of two conjugate diameters, are to each other as the rectangles of the segments into which they divide the diameter.*

*Scholium* 4. This property enables us to describe an ellipse by points when we know two conjugate diameters and the angle which they form with each other.

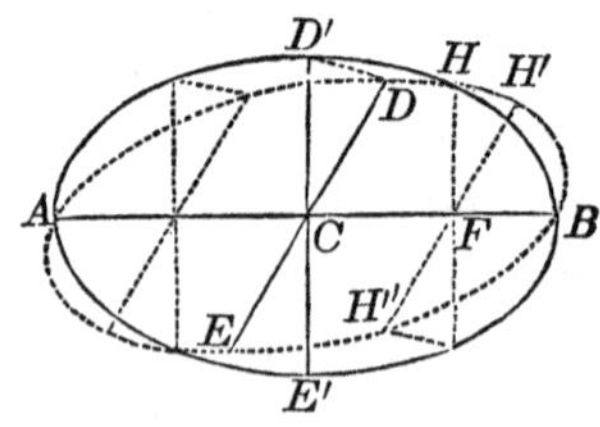


Let $AB$, $ED$, be two conjugate diameters. Turn $ED$ round the centre $C$, until it becomes perpendicular to $AB$, and then describe an ellipse on $AB$ and $E'D'$ as axes.

Take any ordinate to the transverse axis as $FH$, and incline it until it becomes parallel to $CD$: the points $H'$, $H''$, will be on the ellipse described on the conjugate diameters $AB$, $ED$. In a similar manner any number of points may be found.

*Scholium* 5. To determine whether there are any conjugate diameters that are equal to each other, we make $A' = B'$ in the equations

$$A'^2 = \frac{A^2B^2}{A^2 \sin^2 \alpha + B^2 \cos^2 \alpha}, \quad B'^2 = \frac{A^2B^2}{A^2 \sin^2 \alpha' + B^2 \cos^2 \alpha'}:$$

this will give

$$A^2 \sin^2 \alpha' + B^2 \cos^2 \alpha' = A^2 \sin^2 \alpha + B^2 \cos^2 \alpha.$$

If there are any unequal values of $\alpha$ and $\alpha'$ which will satisfy this equation, such values will give equal conjugate diameters.

If we substitute for $\sin^2 \alpha'$, $1 - \cos^2 \alpha'$, and for $\sin^2 \alpha$, $1 - \cos^2 \alpha$, we shall have,

$$A^2 - A^2 \cos^2 \alpha' + B^2 \cos^2 \alpha' = A^2 - A^2 \cos^2 \alpha + B^2 \cos^2 \alpha,$$

or $$(A^2 - B^2) \cos^2 \alpha' = (A^2 - B^2) \cos^2 \alpha;$$

hence, $$\cos \alpha' = - \cos \alpha,$$

or $$\alpha' + \alpha = 180^\circ.$$

Hence, 1st. *The equal conjugate diameters are parallel to the supplementary chords joining the vertices of the axes* (Prop. IV, Sch. 3).

2d. *They form with each other the greatest angle which can be contained by conjugate diameters.*

*Scholium* 6. The *parameter* of any diameter is a third proportional to the diameter and its conjugate. Thus, if $P$ designate the parameter of the diameter $2A'$, we shall have,

$$2A' : 2B' :: 2B' : P,$$

or

$$P = \frac{2B'^2}{A'}.$$

*Scholium* 7. The parameter of the transverse axis is equal to $\frac{2B^2}{A}$; and of the conjugate axis to $\frac{2A^2}{B}$.

It is easily shown, that the chord drawn through the focus, and perpendicular to the transverse axis, is equal to the parameter of that axis.

*Scholium* 8. If through the extremities of either of the conjugate diameters two supplementary chords be drawn, they will enjoy properties analogous to those of the supplementary chords drawn through the vertices of either axis.

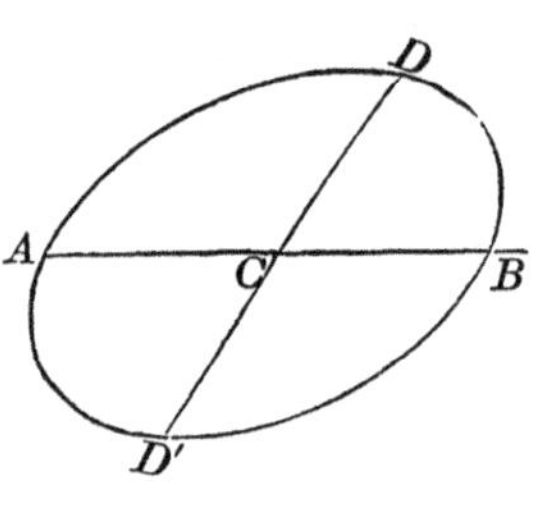


Let $AB$ be the conjugate diameter, which is designated by $2A'$. This line will represent the axis of $X'$, and $DD'$ will correspond to $Y'$. Designate the angle $DCB$ by $\beta$. Then, if through $B$, whose co-ordinates are $y'=0$ and $x'=A'$, a right line be drawn, making with $AB$ an angle equal to $\alpha$, its equation will be of the form (Bk. II, Prop. IV),

$$y = a(x - A'), \quad \text{and} \quad a = \frac{\sin \alpha}{\sin(\beta - \alpha)}.$$

If through $A$, whose co-ordinates are $y'=0$, $x'=-A'$ a second right line be drawn, making with $AB$ an angle $\alpha'$ we shall have,

$$y = a'(x + A'), \qquad a' = \frac{\sin \alpha'}{\sin(\beta - \alpha')},$$

Combining these equations with each other, and with the equation of the ellipse

$$A'^2y^2 + B'^2x^2 = A'^2B'^2,$$

we obtain $$A'^2aa' + B'^2 = 0$$

for the equation of condition when the lines are supplementary chords. In this equation, $a$ and $a'$ express the ratio of the sines of the angles which the supplementary chords make with the conjugate diameters.

PROPOSITION X. PROBLEM.

*To find the equation of a tangent line to an ellipse referred to its conjugate diameters.*

If the co-ordinates of the point of contact be designated by $x''$, $y''$, the equation of the tangent will be of the form

$$y - y'' = a(x - x''),$$

in which it is required to determine the constant quantity $a$. To do this, combine the last equation with the equations

$$A'^2y^2 + B'^2x^2 = A'^2B'^2$$
$$A'^2y''^2 + B'^2x''^2 = A'^2B'^2$$

as in Prop. V; and we shall find

$$a = -\frac{B'^2x''}{A'^2y''},$$

and the equation of the tangent will become

$$y - y'' = -\frac{B'^2x''}{A'^2y''}(x - x'').$$

or by reducing $A'^2yy'' + B'^2xx'' = A'^2B'^2$,

which is of the same form as in the case of the axes.

*Scholium* 1. If a line be drawn through the origin of co-ordinates, its equation will be of the form

$$y = a'x.$$

If this line passes through the point of tangency, the equation will become

$$y'' = a'x'', \quad \text{whence,} \quad a' = \frac{y''}{x''}.$$

If this value of $a'$ be multiplied by that of $a$, in the equation of the tangent, we have

$$aa' = -\frac{B'^2}{A'^2}.$$

By comparing this equation with that deduced in the eighth scholium of the previous proposition, we see, that the same relations exist between the angles which a tangent and the diameter passing through the point of contact, form with the axis of $X'$, as exist between the angles which the supplementary chords form with the same axis.

Therefore, if we make

$$a = a,$$

we shall have

$$a' = a';$$

that is, if the tangent be parallel to one of the supplementary chords, the diameter passing through the point of contact will be parallel to the other: and conversely, if the diameter is parallel to one of the supplementary chords, the tangent at the vertex will be parallel to the other.

This is the same property, with respect to any diameter, as was proved in Prop. VII, with respect to the transverse axis.

*Scholium* 2. This property enables us to draw a tangent line to an ellipse, at a given point, without knowing the axes.

Let $P$ be the point of the curve at which the tangent is to be drawn.

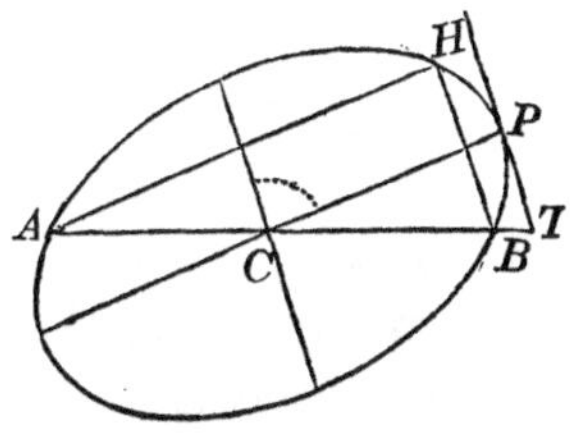


Find the centre $C$ of the ellipse, as in Scholium 2 of the last proposition: draw $CP$, and a diameter $ACB$. Draw $AH$ parallel to $CP$, and join $H$ and $B$. Then draw $PT$ parallel to $HB$ and it will be tangent to the ellipse at $P$.

*Scholium* 3. The same property also enables us to find two conjugate diameters which shall make a given angle with each other.

Let $AB$ be any diameter of the ellipse, and $C$ the given angle which the required conjugate diameters are to make with each other.

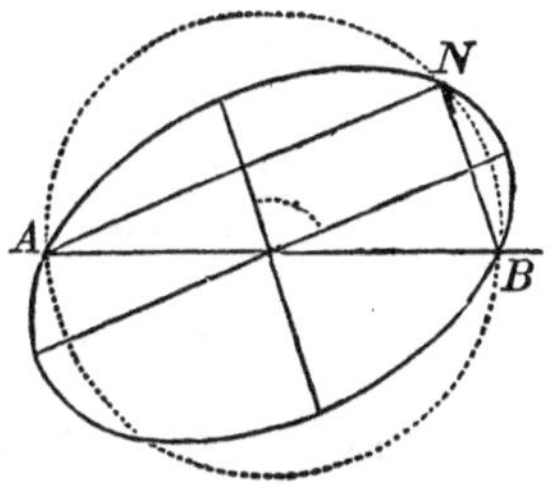


On $AB$, describe an arc capable of containing the given angle $C$ (Geom. Bk. III, Prob. XVI). Through the point $N$, in which this arc intersects the ellipse, draw the supplementary chords $AN$, $BN$. Through the centre, draw two diameters respectively parallel to these chords, and they will be conjugate diameters, and make with each other the given angle.

If the given angle is equal to 90°, the conjugate diameters which are found will be the axes of the ellipse.

If the angle is greater than the angle formed by the supplementary chords drawn through the vertices of the axes, the problem will be impossible, and the circumference of the circle will intersect the ellipse only at the points $A$ and $B$.

9

### PROPOSITION XI. PROBLEM.

*To find the equations which shall express the relation between the values of the axes of an ellipse, and the values of any two conjugate diameters.*

The equation of the ellipse referred to its conjugate diameters, is of the form

$$A'^2y'^2 + B'^2x'^2 = A'^2B'^2;$$

and the formulas for passing from oblique to rectangular axes, the origin remaining the same, are (Bk. II, Prop. XII),

$$x' = \frac{x \sin \alpha' - y \cos \alpha'}{\sin (\alpha' - \alpha)} \qquad y' = \frac{y \cos \alpha - x \sin \alpha}{\sin (\alpha' - \alpha)};$$

Substituting these values of $y'$, $x'$, we have

$$(A'^2 \cos^2\alpha + B'^2 \cos^2\alpha')y^2 + (A'^2 \sin^2\alpha + B'^2 \sin^2\alpha')x^2$$
$$-2(A'^2 \sin \alpha \cos \alpha + B'^2 \sin \alpha' \cos \alpha')xy = A'^2B'^2 \sin^2(\alpha' - \alpha),$$

for the equation of the ellipse referred to its centre and axes

But $\alpha$ and $\alpha'$ must have such values as to make this equation of the known form,

$$A^2y^2 + B^2x^2 = A^2B^2.$$

We must therefore have,

$$A'^2 \cos^2\alpha + B'^2 \cos^2\alpha' = A^2,$$

$$A'^2 \sin^2\alpha + B'^2 \sin^2\alpha' = B^2,$$

$$A'^2 \sin \alpha \cos \alpha + B'^2 \sin \alpha' \cos \alpha = 0$$

$$A'^2B'^2 \sin^2 (\alpha' - \alpha) = A^2B^2,$$

or

$$A'B' \sin (\alpha' - \alpha) = AB.$$

If we add the first and second equations, we shall obtain

$$A'^2 + B'^2 = A^2 + B^2.$$

If to the two last equations, we unite the equation

$$A^2 \text{ tang } \alpha \text{ tang } \alpha' + B^2 = 0,$$

which determines the position of the conjugate diameters with respect to the axes, we shall have the three following equations:

$$A^2 \text{ tang } \alpha \text{ tang } \alpha' + B^2 = 0, \qquad (1)$$

$$A'B' \sin (\alpha' - \alpha) \quad = AB, \qquad (2)$$

$$A'^2 + B'^2 \quad = A^2 + B^2, \quad (3)$$

which express the relations that exist between the axes of an ellipse and any two of the conjugate diameters.

*Scholium* 1. These equations contain six quantities:

$$A, \quad B, \quad A', \quad B', \quad \alpha, \quad \alpha'.$$

If three of these six quantities are known, or given, there will be but three unknown quantities entering into the three equations; the values of these may, therefore, be found from the equations.

*Scholium* 2. If we know the angle which the conjugate diameters make with each other, it will be equivalent to knowing $\alpha$ or $\alpha'$ For, designate the known angle by $\beta$;

then, $\alpha' - \alpha = \beta,$

or $\alpha' = \beta + \alpha;$

hence, $\text{tang } \alpha' = \text{tang } (\beta + \alpha) = \dfrac{\text{tang } \beta + \text{tang } \alpha}{1 - \text{tang } \beta \text{ tang } \alpha}.$

Substituting this value of tang $\alpha'$ in the first equation, and we have

$$A^2 \text{ tang}^2 \alpha + (A^2 - B^2) \text{ tang } \alpha \text{ tang } \beta + B^2 = 0 ;$$

from which we can find tang $\alpha$, and consequently $\alpha$, in terms of the axes and the known quantity, tang $\beta$.

*Scholium* 3. To interpret the second equation,

$$A'B' \sin (\alpha' - \alpha) = AB,$$

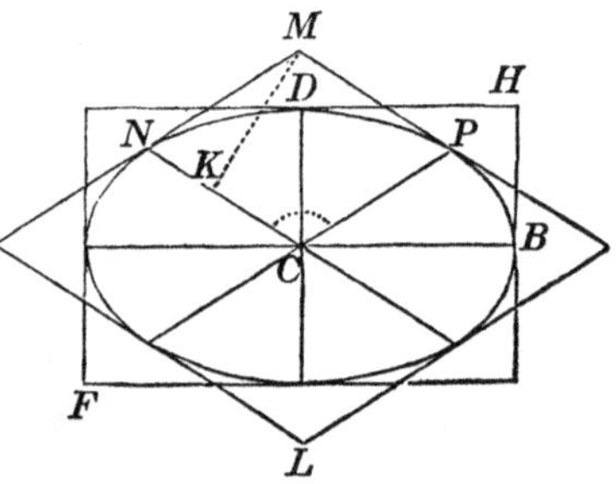


let us suppose the ellipse whose centre is $C$, to be circumscribed by a rectangle formed by drawing tangents at the vertices of the axes, and also by a parallelogram formed by drawing tangents at the vertices of the conjugate diameters, which are represented by $2A'$ and $2B'$.

From $M$ draw $MK$ perpendicular to $CN$. The angle $NCP$ is designated by $\alpha' - \alpha$, and since $MNC$ is the supplement of $NCP$, its sine will be equal to $\sin (\alpha' - \alpha)$. Further, $NM = CP = A'$. Therefore (Trig. Th. I, Cor.),

$$MK = A' \sin (\alpha' - \alpha).$$

Hence, $\quad A'B' \sin (\alpha' - \alpha) = CPMN \quad$ (Mens. Prob. I)

The second member, $A \times B$, is equal to the rectangle $CBHD$. But the parallelogram $CPMN$ is one-fourth the parallelogram $ML$, and the rectangle $CBHD$ is one-fourth the rectangle $HF$: hence, the second equation expresses the following property:

*The rectangle which is formed by drawing tangent lines through the vertices of the axes, is equivalent to the paral-*

*lelogram which is formed by drawing tangents through the vertices of conjugate diameters.*

*Scholium* 4. To interpret the third equation,

$$A'^2 + B'^2 = A^2 + B^2,$$

we first multiply both members by 4, and obtain,

$$4A'^2 + 4B'^2 = 4A^2 + 4B^2,$$

which expresses the following property.

*The sum of the squares described on the axes of an ellipse is equal to the sum of the squares described on two conjugate diameters.*

## *Of the Polar Equation of the Ellipse and the Measure of its Surface.*

### PROPOSITION XII. PROBLEM.

*To find the general polar equation of the ellipse.*

If we designate the co-ordinates of the pole $P$, by $a$ and $b$, and estimate the angles $v$ from the line $PX'$ parallel to the transverse axis, we shall have the following formulas for passing from rectangular to polar co-ordinates (Bk. II, Prop. XV): **viz.,**

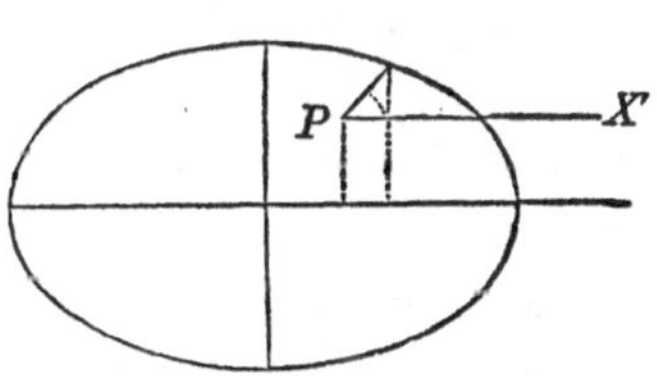


$$x = a + r\cos v, \qquad y = b + r\sin v.$$

If we substitute these values of $x$ and $y$ in the equation,

$$A^2y^2 + B^2x^2 = A^2B^2,$$

we obtain

$$\left.\begin{array}{l} A^2 \sin^2 v \\ +B^2 \cos^2 v \end{array}\right| r^2 + \left.\begin{array}{l} 2A^2 b \sin v \\ +2B^2 a \cos v \end{array}\right| r + A^2 b^2 + B^2 a^2 - A^2 B^2 = 0$$

which is the general polar equation of the ellipse.

*Scholium* 1. If we suppose the pole $P$ to be placed on the ellipse, its co-ordinates $a$ and $b$ will satisfy the equation of the ellipse, and give

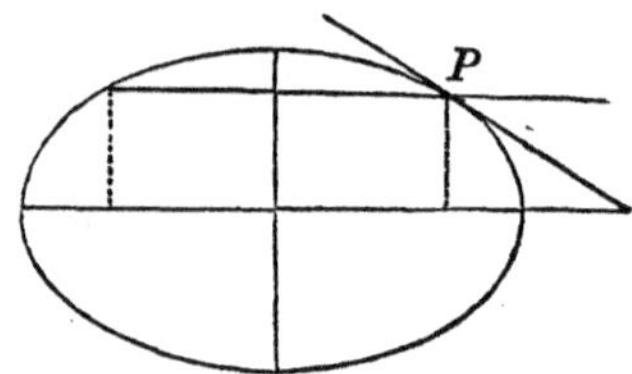


$$A^2 b^2 + B^2 a^2 - A^2 B^2 = 0.$$

The polar equation will then reduce to

$$\left.\begin{array}{l} A^2 \sin^2 v \\ +B^2 \cos^2 v \end{array}\right| r^2 + \left.\begin{array}{l} 2A^2 b \sin v \\ +2B^2 a \cos v \end{array}\right| r \Big\} = 0;$$

which becomes

$$\left.\begin{array}{l} A^2 \sin^2 v \\ +B^2 \cos^2 v \end{array}\right| r + \left.\begin{array}{l} 2A^2 b \sin v \\ +2B^2 a \cos v \end{array}\right\} = 0,$$

by dividing by that value of $r$ which is 0.

The value of $r$, in this equation, represents the distance from $P$ to the second point in which the radius-vector meets the curve.

If we make the second value of $r$ also equal to 0, the two points in which the radius-vector intersects the ellipse will unite, and the radius-vector will become tangent to the curve. Making $r=0$, we obtain

$$2A^2 b \sin v + 2B^2 a \cos v = 0;$$

hence, $$\tan v = -\frac{B^2 a}{A^2 b},$$

which represents, as it ought to do, the tangent of the angle which the tangent line makes with the axis of abscissas.

*Scholium* 2. If the pole be placed at the centre of the ellipse, its co-ordinates will become

$$a = 0, \qquad b = 0.$$

Substituting these values in the general polar equation, and it will give

$$r = \frac{AB}{\sqrt{A^2 \sin^2 v + B^2 \cos^2 v}}.$$

Making $v = 0$, we have

$$r = A.$$

If we make $v = 90^\circ$, we have

$$r = B.$$

If we suppose $A = B$, the ellipse will become a circle, and we shall have

$$r = A$$

for all values of $v$.

*Scholium* 3. Let us now discuss the polar equation when the pole is placed at either of the foci.

The co-ordinates of the focus $F$, are

$$b = 0, \qquad a = +\sqrt{A^2 - B^2}.$$

If these values be substituted in the equation

$$\left.\begin{matrix} A^2 \sin^2 v \\ + B^2 \cos^2 v \end{matrix}\right| r^2 + \left.\begin{matrix} 2A^2 b \sin v \\ + 2B^2 a \cos v \end{matrix}\right| r + A^2 b^2 + B^2 a^2 - A^2 B^2 = 0,$$

it reduces to

$$(A^2 \sin^2 v + B^2 \cos^2 v) r^2 + 2B^2 a \cos v . r = B^4$$

Resolving the equation with refeience to $r$, and the radical part becomes

$$B^4(A^2\sin^2 v+B^2\cos^2 v)+B^4a^2\cos^2 v,$$

which, by placing for $a^2$ its value $A^2-B^2$, reduces to

$$A^2B^4(\sin^2 v+\cos^2 v),\qquad\text{or}\quad A^2B^4\,;$$

which gives two rational values,

$$r=-\frac{B^2(a\cos v-A)}{A^2\sin^2 v+B^2\cos^2 v},\ \text{and}\ r=-\frac{B^2(a\cos v+A)}{A^2\sin^2 v+B^2\cos^2 v}$$

These values may be placed under more simple forms: for we have, by substituting for $\sin^2 v$, $1-\cos^2 v$,

$$A^2\sin^2 v+B^2\cos^2 v=A^2-(A^2-B^2)\cos^2 v=A^2-a^2\cos^2 v.$$

But we also have

$$A^2-a^2\cos^2 v=(A+a\cos v)(A-a\cos v),$$

which being substituted in the values of $r$, will give

$$r=\frac{B^2}{A+a\cos v},\ \text{and}\quad r=-\frac{B^2}{A-a\cos v}.$$

Now, the second value of $r$ is constantly negative for all values of $v$. For, the numerator $B^2$ is positive, and in the denominator $a<A$ and $\cos v<1$; hence, $A-a\cos v$ is positive; therefore, the essential sign of the second member is the same as its algebraic sign, that is, negative. Hence, this value must be rejected.

In regard to the first value,

$$r=\frac{B^2}{A+a\cos v},$$

it will give points of the curve from $v=0$ to $v=360^\circ$, which will determine the entire ellipse. In this equation the values of $r$ begin at the vertex *nearest* the pole.

*Scholium* 4. The pole can be removed to the focus $F'$ by simply changing the sign of $a$. We should then have

$$r = \frac{B^2}{A - a\cos v},$$

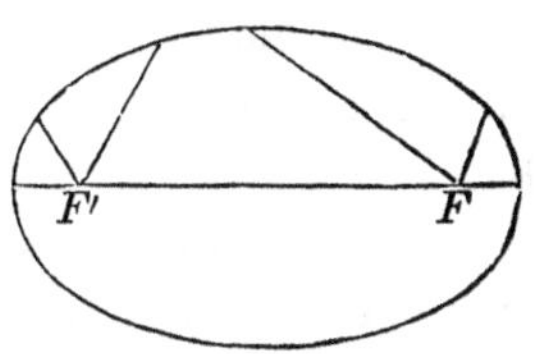


in which the values of $r$ begin at the vertex *farthest* from the pole, and $v$ increases in the same direction as before, from 0 to 360°.

*Scholium* 5. It simplifies the polar equation to introduce into it the eccentricity (Prop. I, Sch. 4).

Designate the eccentricity by $e$; then,

$$\frac{a}{A} = e, \qquad \text{or} \qquad a = Ae$$

Also $$\frac{A^2 - B^2}{A^2} = e^2, \quad \text{hence,} \quad B^2 = A^2(1 - e^2).$$

Substituting these values in the two last equations, they become

$$r = \frac{A(1 - e^2)}{1 + e\cos v}, \quad \text{and} \quad r = \frac{A(1 - e^2)}{1 - e\cos v}.$$

*Scholium* 6. It should be remarked, that the common numerator in these values of $r$, is equal to half the parameter of the transverse axis.

For $$A(1 - e^2) = A\left(1 - \frac{(A^2 - B^2)}{A^2}\right) = \frac{B^2}{A}.$$

PROPOSITION XIII. THEOREM.

*The area of an ellipse is equal to the product of its semi-axes multiplied by the circumference of a circle whose diameter is unity*

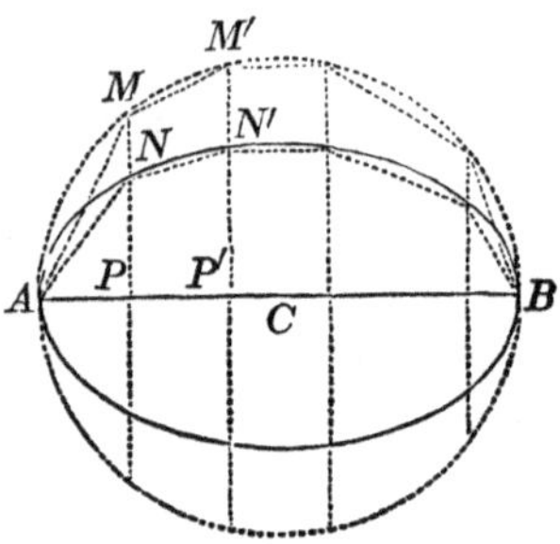


Let $AB$ be the transverse axis of an ellipse and $C$ the centre.

On the transverse axis $AB$ describe the circumference of a circle, then, if $Y$ designate the ordinates of the circle, and $y$ the corresponding ordinates of the ellipse, we shall have (Prop. III),

$$Y : y :: A : B; \quad \text{and}$$

$$Y' : y' :: A : B; \quad \text{hence,}$$

$$Y + Y' : y + y' :: A : B$$

(Geom. Bk. II, Prop. X), and the areas of the circle and ellipse are to each other in the same ratio.

To prove this, inscribe a polygon in the circle and from the angular points draw ordinates to the diameter $AB$. Join the points in which these ordinates intersect the ellipse, and there will be inscribed in the ellipse a polygon of an equal number of sides.

We shall then have, for one of the trapezoids $PNN'P'$ of the ellipse,

$$PP'\left(\frac{PN + P'N'}{2}\right) \quad \text{or} \quad (x - x')\left(\frac{y + y'}{2}\right);$$

and for the corresponding trapezoids $PMM'P'$ of the circle

$$PP'\left(\frac{PM + P'M'}{2}\right) \quad \text{or} \quad (x - x')\left(\frac{Y + Y'}{2}\right).$$

These trapezoids are to each other as

$$Y + Y' \quad \text{to} \quad y + y', \quad \text{that is, as} \quad A \quad \text{to} \quad B,$$

and the entire surfaces of the polygons of which the trapezoids are like parts, will be to each other in the same ratio. As this will hold true whatever may be the number of sides of the polygons, it will be true when that number is indefinitely increased, in which case one of the polygons will become the circle and the other the ellipse.

Let us designate the area of the circle by $S$, and the area of the ellipse by $s$, we shall then have,

$$S : s :: A : B,$$

that is, the area of the circle is to the area of the ellipse as $A$ to $B$, and consequently,

$$\frac{s}{S} = \frac{B}{A},$$

But the area of the circle is equal to $A^2\pi$, (Geom. Bk. V, Prop. XII, Cor. 2): we shall then have for the area of the ellipse

$$s = AB\pi.$$

*Corollary.* The area of the ellipse is a mean proportional between the two circles described on the axes. For, the area of the circle described on the transverse axis is $A^2\pi$; and the area of that on the conjugate axis is $B^2\pi$: their product is $A^2B^2\pi^2$, the square root of which is $AB\pi$

# BOOK V.

## *Of the Parabola.*

The *parabola* is a curve of which any point is equally distant from a fixed point and a given straight line.

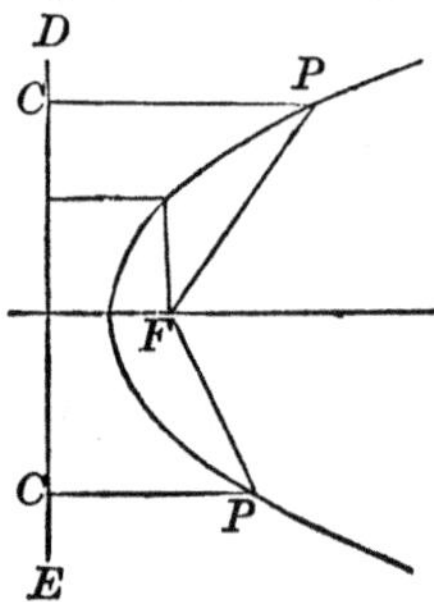


The fixed point is called the *focus* of the parabola, and the given straight line is called the *directrix.*

Thus, if $F$ be the fixed point and $ED$ the given line, and the point $P$ be so moved that $PF$ shall be constantly equal to $PC$, the point $P$ will describe a parabola of which $F$ is the focus and $DE$ the directrix.

### PROPOSITION I. PROBLEM.

*To find the equation of the parabola.*

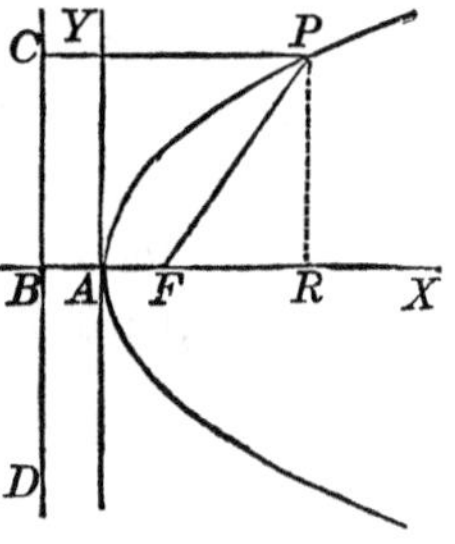


Let $F$ be the focus and $DC$ the directrix. Denote the distance $FB$, from the focus to the directrix, by $p$, and let the point $A$, equally distant from $B$ and $F$, be assumed as the origin of a system of rectangular co-ordinates of which $AX$, $AY$, are the axes. The distance $AF$ will be equal to $\frac{p}{2}$.

Let $x$ and $y$ be the co-ordinates of any point of the curve as $P$: then will the distance $FP$, from this point to $F$, or which the co-ordinates are $x' = \frac{p}{2}$ and $y' = 0$, be expressed by

$$\sqrt{y^2 + \left(x - \frac{p}{2}\right)^2}.$$

But by the definition of the curve,

$$FP = PC = BA + AR = \frac{p}{2} + x.$$

Hence,

$$\sqrt{y^2 + \left(x - \frac{p}{2}\right)^2} = \frac{p}{2} + x,$$

or, $$y^2 + x^2 - px + \frac{p^2}{4} = \frac{p^2}{4} + px + x^2,$$

hence, $$y^2 = 2px,$$

which is the equation of the parabola referred to the rectangular axes of which $A$ is the origin.

*Scholium* 1. The axis of abscissas $AX$ is called the *axis* of the parabola, and the origin $A$ is called the *vertex* of the axis, or principal *vertex*; and the constant quantity $2p$ is called the *parameter*.

The equation of the parabola gives

$$y = \pm \sqrt{2px},$$

from which we see, that for every value of $x$ there will be two equal values of $y$ with contrary signs. Hence, *the parabola is symmetrical with respect to its axis.*

We see further, that $y$ will increase with $x$, and will have real values so long as $x$ is positive. Hence, the curve extends indefinitely in the direction of $x$ positive.

If we make $x = 0$, we have,

$$y = \pm 0,$$

which shows that the axis of $y$ is tangent to the curve at the origin.

If we make $x$ negative, we shall have,

$$y = \pm\sqrt{-2px},$$

or, $y$ imaginary; which shows, that *the curve does not pass the axis of* Y *and extend on the side of* x *negative.*

*Scholium* 2. In the ellipse the parameter of the transverse axis is a third proportional to the axes : in the parabola it is a third proportional to any abscissa and the corresponding ordinate. For, from the equation

$$y^2 = 2px,$$

we have,

$$x : y :: y : 2p.$$

*Scholium* 3. If the abscissa $x$ be made equal to $\frac{1}{2}p$ we shall have

$$y^2 = p^2, \quad \text{or} \quad y = p, \quad \text{or} \quad 2y = 2p,$$

which shows that *the double ordinate through the focus is equal to the parameter.*

*Scholium* 4. In the parabola, the squares of the ordinates are to each other as the corresponding abscissas. For, designating any two ordinates by $y'$, $y''$, and their corresponding abscissas by $x'$, $x''$, we shall have

$$y'^2 = 2px' \quad \text{and} \quad y''^2 = 2px'';$$

hence, $$y'^2 : y''^2 :: x' : x'',$$

by omitting the common factor $2p$.

*Scholium* 5. By a system of reasoning similar to that in Sch. 5, Prop. I, Bk. IV, we find the conditions for determining the position of a point with respect to the curve. They are,

for a point without the curve, $y^2 - 2px > 0$;

for a point on the curve, $y^2 - 2px = 0$;

for a point within the curve, $y^2 - 2px < 0$.

*Scholium* 6. The properties of the parabola which have been explained, afford an easy method of describing it mechanically.

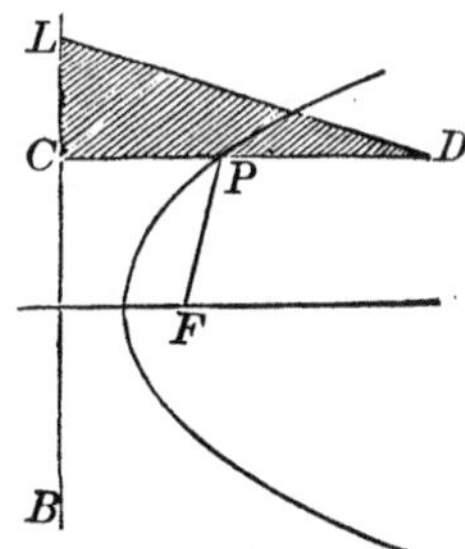


Let $BL$ be a given line, and $LCD$ a triangular ruler, right-angled at $C$. Take a thread, the length of which is equal to the side $CD$, and attach one extremity at $D$, and the other at any point, as $F$. Place a pencil against the thread and the ruler, making tense the parts of the thread $FP$, $PD$. Then if the side $CL$ of the ruler be moved along the line $BL$, the pencil will describe a parabola, of which $F$ is the focus and $BL$ the directrix: for, the distance $PF$ will be equal to $PC$, for every position of the ruler.

*Scholium* 7. If the parameter of the parabola is known we have a simple construction for determining points of the curve.

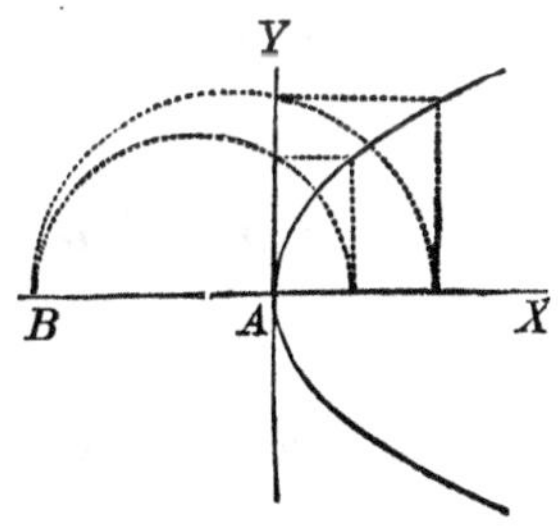


Let $AX$, $AY$ be the co-ordinate axes. The equation of the curve is,

$$y^2 = 2px.$$

From the origin $A$, lay off a distance $AB$, on the negative side abscissas, equal to $2p$. Then from $A$ lay off any distance as

$AP$ and draw $PM$ perpendicular to $AX$. On $BP$ as a diameter, describe a semi-circumference, and through $Q$ where it intersects the axis $AY$, draw $QM$ parallel to $AX$; the point $M$ where it intersects $PM$, will be a point of the curve. For, from the equation of the circle,

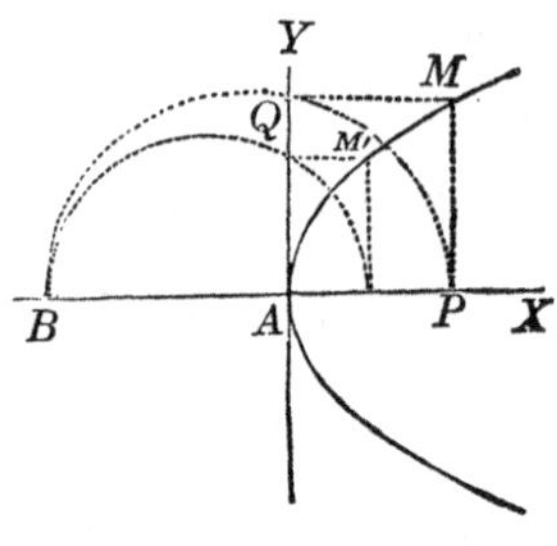


$$\overline{AQ}^2 = BA . AP,$$

hence,
$$y^2 = 2px,$$

for any point, $M$ or $M'$.

PROPOSITION II. PROBLEM.

*To find the equation of a tangent line to the parabola.*

Let us designate the co-ordinates of the point of tangency $P$ by $x''$, $y''$; the equation of a straight line passing through this point, will be

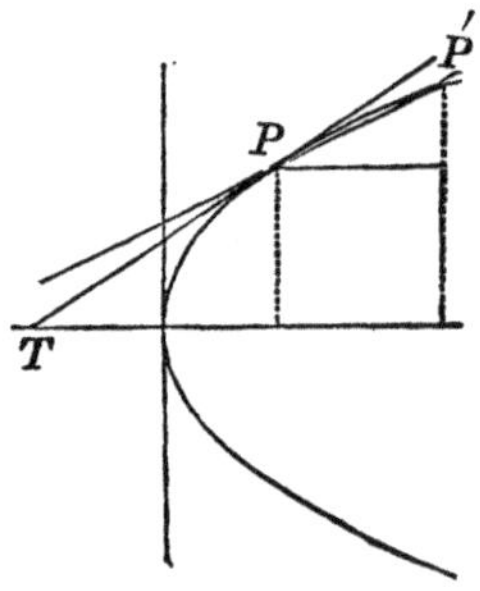


$$y - y'' = a(x - x'');$$

it is required to determine $a$ when the right line is tangent to the parabola. The equation of the parabola is

$$y^2 = 2px \ ;$$

and since the point of tangency is on the curve, we also have

$$y''^2 = 2px''.$$

Subtracting the last equation from the preceding, we obtain

$$(y + y'')(y - y'') = 2p(x - x'').$$

Combining this with the equation of the line passing through $P$, and we have

$$(y+y'')a(x-x'')=2p(x-x''),$$

or $$a(y+y'')=2p,$$

in which $y$ is the ordinate of $P'$.

Making $y=y''$, we have

$$a=\frac{p}{y''}.$$

Substituting this value in the equation of the line passing through $P$, and we have

$$y-y''=\frac{p}{y''}(x-x''),$$

or by reducing, and observing that $y''^2=2px''$,

$$yy''=p(x+x''),$$

which is the equation of the tangent line.

*Scholium* 1. It is easily proved that every point of the tangent line, except the point of contact, lies without the curve.

For, subtracting twice the last equation from

$$y''^2=2px'',$$

and then adding $y^2$ to both members, we obtain

$$y''^2-2yy''+y^2=y^2-2px,$$

or $$(y''-y)^2=y^2-2px.$$

But the first member being a square, is positive; hence, the second member will be positive; therefore, the points whose co-ordinates are $x$ and $y$, are without the curve (Prop. I; Sch. 5).

If the co-ordinates of the point, at which the tangent touches the curve, be substituted in the equation, both members will reduce to 0, which shows that the point of tangency is on the curve.

*Scholium* 2. If in the equation of the tangent

$$yy'' = p\,(x + x''),$$

we make $y = 0$, we shall have

$$0 = p(x + x'').$$

But since the factor $p$ is a constant quantity, we must have,

$$x + x'' = 0\,;$$

that is, $\quad x'' = AD = -x = -AT.$

That is, *the sub-tangent is double the abscissa*, or *is bisected at the vertex.*

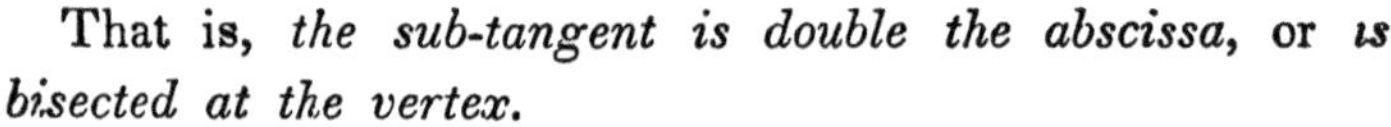

The analytical condition expressed by

$$x + x'' = 0, \qquad \text{or} \qquad AT + AD = 0,$$

indicates that the quantities are equal to each other, and estimated on different sides of the origin of co-ordinates.

*Scholium* 3. This property indicates a simple method of drawing a tangent line to the curve when the point of contact $P$, and the axis, are given.

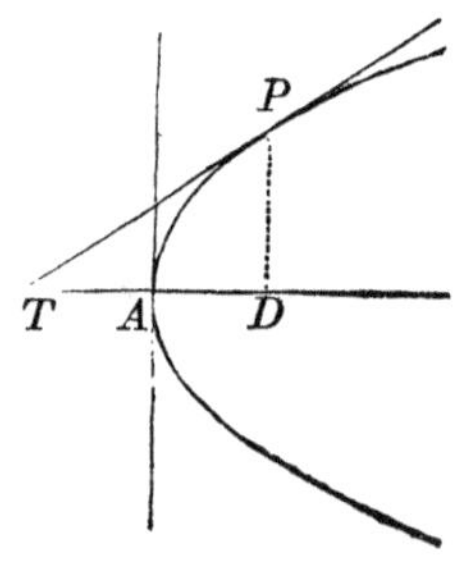


For, from $P$ draw $PD$ perpendicular to the axis, and then make $AT = AD$. Draw a right line through $T$ and $P$, and it will be the tangent required.

## PROPOSITION III. PROBLEM.

*To find the equation of a normal line to the parabola.*

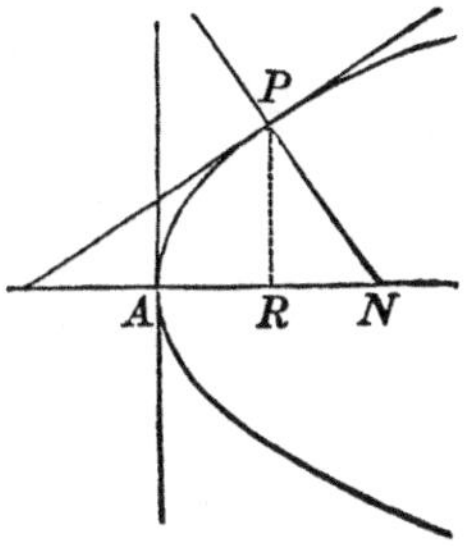

Let $x''$, $y''$ be the co-ordinates of he point of tangency. Then, the equation of the normal will be of the form

$$y-y''=a'(x-x''),$$

and since it is perpendicular to the tangent

$$aa'+1=0.$$

But we have already found

$$a=\frac{p}{y''}, \qquad \text{hence} \qquad a'=-\frac{y''}{p};$$

therefore, we have

$$y-y''=-\frac{y''}{p}(x-x'')$$

for the equation of the normal.

*Scholium* 1. If in the equation of the normal

$$y-y''=-\frac{y''}{p}(x-x''),$$

we make $y=0$, and then find the value of $x-x''$, we shall have,

$$x-x''=p.$$

But $x$ is equal to the distance $AN$, and $x''$ to the distance $AR$: hence, $x-x''=p$ is equal to $RN$: that is, *the sub-normal is constant, and equal to half the parameter.*

*Scholium* 2. This property furnishes an easy method of drawing a tangent line to the parabola, at a given point $P$.

For, draw the ordinate $PR$ to the axis, and from the foot $R$, lay off a distance $RN = p$, and join $P$ and $N$. Then draw $TP$ perpendicular to $PN$ at $P$, and it will be the tangent required.

### PROPOSITION IV. THEOREM.

*A tangent line to the parabola at any point of the curve makes equal angles with the axis and with the line drawn from the point of tangency to the focus.*

Designate the co-ordinates of the point of tangency $P$ by $x''$, $y''$, and the co-ordinates of the focus $F$, by

$$x' = \frac{p}{2} \quad \text{and} \quad y' = 0.$$

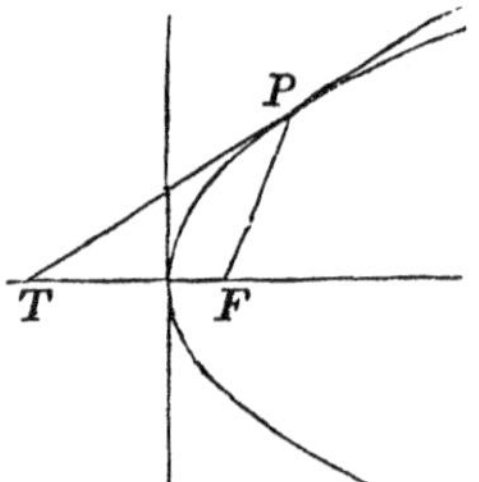


The equation of a straight line passing through the focus will be of the form

$$y - y' = a'(x - x'),$$

and if it passes through the point of tangency, we shall have

$y'' - y' = a'(x'' - x')$: but the co-ordinates of the focus being $x = \frac{p}{2}$ and $y' = 0$, we have, $y'' = a'\left(x'' - \frac{p}{2}\right)$,

$$\text{and} \qquad a' = \frac{y''}{x'' - \frac{p}{2}}.$$

If we designate the angle $TPF$ by $V$, and the tangent of $PTF$ by $a$, we shall have

$$\text{tang } V = \frac{a' - a}{1 + a'a}.$$

But we have before found

$$a = \frac{p}{y''};$$

then substituting for $a'$ and $a$ their values, and observing that $y''^2 = 2px''$, we shall find

$$\text{tang } V = \frac{p}{y''}:$$

hence, the triangle $FTP$ is isosceles.

*Scholium* 1. This property gives an easy construction for drawing a tangent line to the curve at a given point, as $P$.

Join $P$ and the focus $F$: then, lay off from $F$ a distance $FT$ equal to $FP$, and join $P$ and $T$, and $PT$ will be the tangent required.

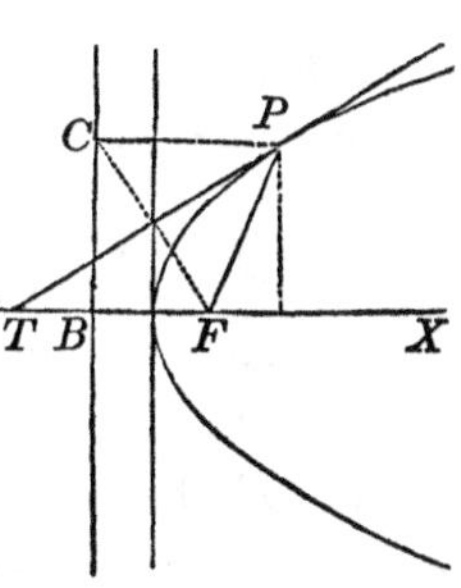


2d *Method.* Draw the directrix $BC$, and from $P$, draw $PC$ perpendicular to it. Draw $FC$, and from $P$ draw $PT$ perpendicular to $FC$, and it will be tangent to the parabola at $P$.

For, since $PC$ and $PF$ are equal, the triangle $PCF$ is isosceles; hence, $PT$ drawn perpendicular to the base $FC$, bisects the vertical angle $CPF$. But because of the parallels $CP$, $TX$, the angle $CPT$ is equal to $PTF$: hence, $PTF$ is equal to $TPF$, and therefore $TP$ is tangent to the parabola at $P$.

*Scholium* 2. We can also draw a tangent that shall be parallel to a given line.

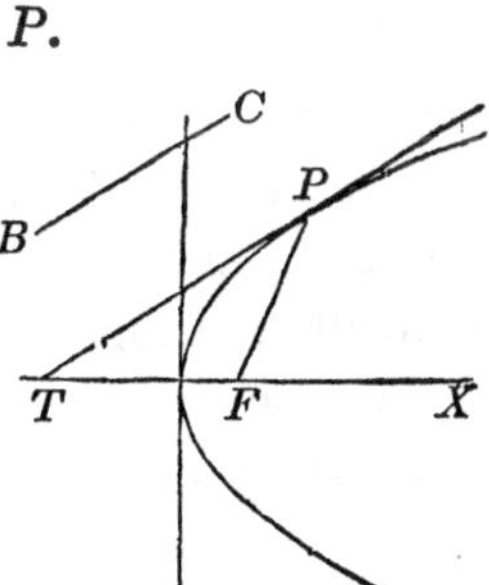


Let $BC$ be the given line. At the focus $F$, lay off an angle $XFP$ equal to twice the angle which the given line makes with the axis of $X$.

Through $P$, the point at which $FP$

intersects the curve, draw $PT$ parallel to $BC$, and it will be the tangent required.

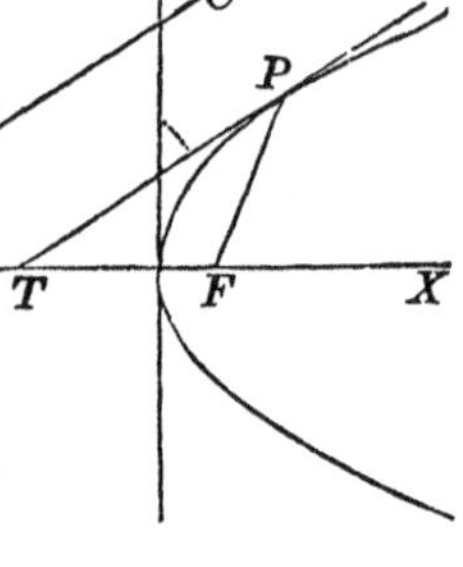


For, the outward angle $PFX$ is equal to the sum of the angles $P$ and $PTF$. But $PTF$ being equal to the angle which $BC$ makes with the axis of $X$, is equal to one-half of $PFX$; hence, the angle at $P$ is equal to half of $PFX$: therefore, the triangle $PTF$ is isosceles, and consequently $PT$ is tangent to the curve at $P$.

*Scholium* 3. The same property also enables us to draw a tangent line to the parabola from a given point without the curve.

Let $G$ be the given point through which the tangent line is to be drawn.

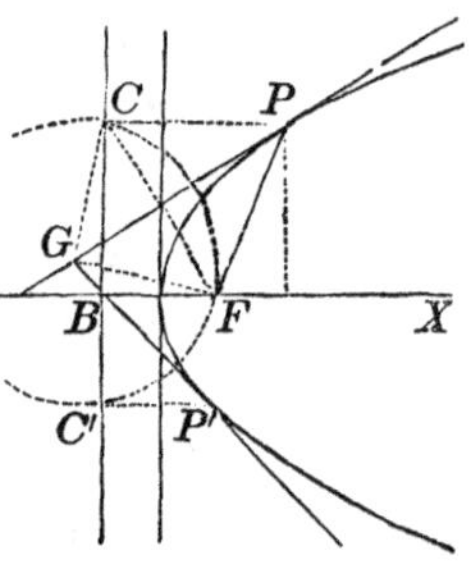


With $G$ as a centre, and a radius equal to $GF$, the distance to the focus, describe the arc of a circle intersecting the directrix at $C$ and $C'$. Through $C$ and $C'$ draw two lines parallel to the axis $BX$, intersecting the parabola in $P$ and $P'$. Through $G$, draw $GP$ and $GP'$, and they will be tangents to the parabola at $P$ and $P'$.

For, join $P$ and the focus $F$. Then, since $P$ is a point of the parabola $PF = PC$; and by construction $GF = GC$: hence, the line $GP$ has two points, $G$ and $P$, equally distant from $C$ and $F$: it is, therefore, perpendicular to $CF$ (Geom. Bk. I, Prop. XVI, Cor.); and is consequently a tangent, by scholium 1.

It may be proved in a similar manner, that $GP'$ is tangent to the curve at $P'$.

*Scholium* 4. It may be shown analytically, that two tangents can be drawn to the parabola from any point without the curve.

Denote the co-ordinates of the given point without the curve by $x'$, $y'$: the equation of the tangent will then become

$$y'y''=p(x'+x'');$$

but since the point of contact is on the curve, we also have

$$y''^2=2px''.$$

Combining these equations, we shall find two real values for $x''$, and two values for $y''$: but these values will be the co-ordinates of the points of contact: hence, there are two points of contact, and therefore, *two tangents may be drawn to the curve from a given point without.*

## PROPOSITION V. THEOREM.

*If from the focus of a parabola a line be drawn perpendicular to a tangent.*

1st. *The point of intersection will be on the axis of* Y · and,

2dly. *The square of the perpendicular will vary as the distance from the focus to the point of contact.*

The co-ordinates of the focus $F$ being, $x'=\frac{1}{2}p$, $y'=0$, the equation of a line passing through it, will be

$$y=a'\left(x-\frac{p}{2}\right).$$

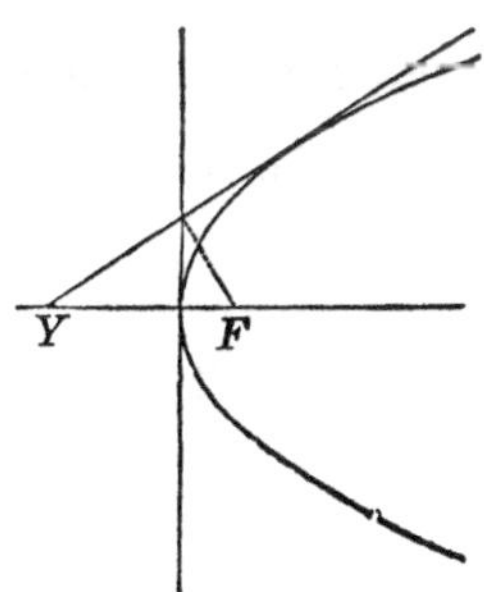


But the condition that this line shall be perpendicular to the tangent gives, $aa' + 1 = 0$, hence,

$$a' = -\frac{1}{a} = -\frac{1}{\frac{p}{y''}} = -\frac{y''}{p}:$$

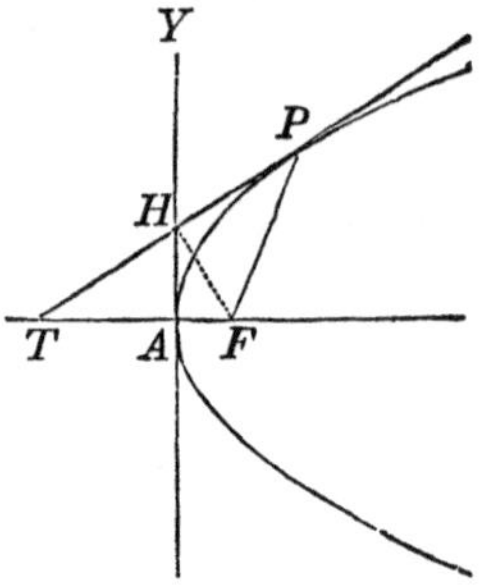

the equation of the perpendicular therefore becomes,

$$y = -\frac{y''}{p}\left(x - \frac{p}{2}\right).$$

Combining this with the equation of the tangent line, which is

$$yy'' = p(x + x''),$$

and substituting for $y''^2$ its value $2px''$, and reducing, we find,

$$x(2x'' + p) = 0;$$

an equation which can only be satisfied when $x = 0$: hence the point $H$ at which the perpendicular meets the tangent is on the axis of $Y$.

To prove the second part of the proposition we have only to consider, that in the right-angled triangle $TFH$,

$$\overline{FH}^2 = FT \times FA, \quad \text{(Geom. Bk. IV, Prop. XXIII)},$$

or, $$\overline{FH}^2 = FP \times FA,$$

But $FA$ is constant, being equal to $\frac{1}{2}p$: hence, $\overline{FH}^2$ varies as the distance $FP$.

## *Of the Parabola referred to Oblique Co-ordinate Axes.*

We have thus far deduced the properties of the parabola from its equation, obtained by referring the curve to a system of rectangular co-ordinates, having their origin at the vertex. We now propose to develop some of the properties of the curve by referring it to a system of oblique co-ordinates.

### PROPOSITION VI. PROBLEM.

*To find a system of co-ordinate axes, such, that the equation of the parabola when referred to them, shall be of the same form as when referred to rectangular axes having their origin at the vertex.*

The formulas for passing from rectangular to oblique axes, are (Bk. II, Prop. XI, Sch.),

$$x = a + x' \cos \alpha + y' \cos \alpha', \qquad y = b + x' \sin \alpha + y' \sin \alpha',$$

in which it is required to find such values for the undetermined constants $a$, $b$, $\alpha$, and $\alpha'$, as to cause the new axes to fulfil the required conditions.

Substituting the values of $x$ and $y$ in the equation

$$y^2 = 2px,$$

and it becomes,

$$\left.\begin{array}{l} y'^2 \sin^2\alpha' + 2x'y' \sin \alpha \sin \alpha' + x'^2 \sin^2\alpha + b^2 - 2ap \\ \quad + 2(b \sin \alpha' - p \cos \alpha')y' + 2(b \sin \alpha - p \cos \alpha)x' \end{array}\right\} = 0.$$

In order that this equation shall be of the form

$$y^2 = 2px,$$

we must give such values tc the undetermined constants as shall make,

$$b^2 - 2ap = 0 \quad (1)$$
$$\sin^2\alpha = 0 \quad (2)$$
$$\sin\alpha \sin\alpha' = 0 \quad (3)$$
$$b\sin\alpha' - p\cos\alpha' = 0 \quad (4);$$

and we are at liberty to assign four arbitrary conditions, since there are four undetermined quantities. Having introduced these conditions, the equation becomes,

$$y'^2 = \frac{2p}{\sin^2\alpha'}x'.$$

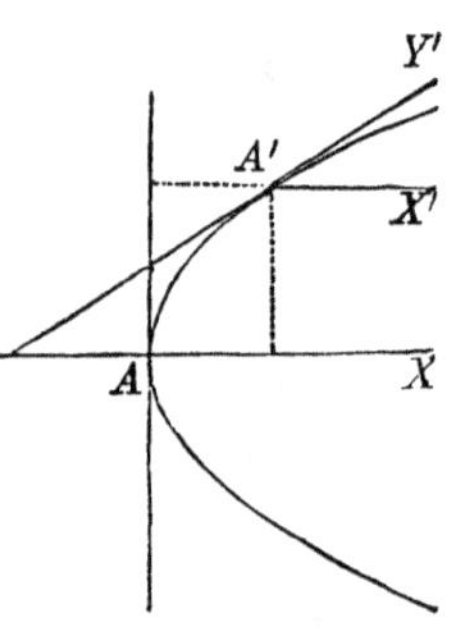


Let us now consider these four conditions separately.

The first equation of condition is of the same form as the equation of the parabola referred to the primitive axes. Therefore, the co-ordinates of the new origin will satisfy the primitive equation, and hence the new origin is on the curve at some point as $A'$.

In the second equation of condition, we have,

$$\sin^2\alpha = 0, \qquad \text{hence,} \qquad \alpha = 0,$$

which shows, that the new axis of abscissas $X'$, is parallel to the primitive axis $AX$.

The third equation of condition

$$\sin\alpha \sin\alpha' = 0,$$

is satisfied by virtue of the $\sin\alpha = 0$; hence it is nothing more than the second.

The fourth equation

$$b\sin\alpha' - p\cos\alpha' = 0,$$

gives, $\qquad \tan\alpha' = \dfrac{p}{b},$

and since this value of tang $\alpha'$ is the same as that found in Prop. II, for the tangent of the angle which the tangent line makes with the axis of $X$, we conclude, that the new axis $Y'$ is tangent to the parabola at the new origin $A'$.

*Scholium* 1. Let us now resume the equation,

$$y'^2 = \frac{2p}{\sin^2\alpha'}x',$$

and to simplify the form of it, put

$$\frac{2p}{\sin^2\alpha'} = 2p';$$

we shall then have, by omitting the accents of the variables,

$$y^2 = 2p'x,$$

for the equation of the parabola referred to the new axes.

In this equation, every value of $x$ will give two equal values of $y$ with contrary signs: hence, the curve is symmetrical with respect to the axis $A'X'$; or, *this axis bisects all chords of the parabola which are parallel to the tangent* A′Y′.

The term *diameter*, designates any straight line which bisects a system of chords drawn parallel to each other and terminating in the curve; and the curve is said to be *symmetrical* with respect to the diameter, whether the chords are oblique or perpendicular to it. In this sense, therefore, every line drawn parallel to the axis $AX$, is a diameter of the parabola: hence, *all diameters of the parabola are parallel to each other, a property which shows that the centre of the curve is at an infinite distance from the vertex.*

*Scholium* 2. In the equation

$$y'^2 = \frac{2p}{\sin^2\alpha'}x'$$

or in the reduced equation

$$y^2 = 2p'x,$$

$$\frac{2p}{\sin^2 \alpha'}, \quad \text{or its equal} \quad 2p',$$

is called the *parameter* of the new diameter $A'X'$.

PROPOSITION VII. THEOREM.

*The parameter of any diameter is equal to four times the distance from the vertex of that diameter to the directrix or four times the distance from the vertex to the focus.*

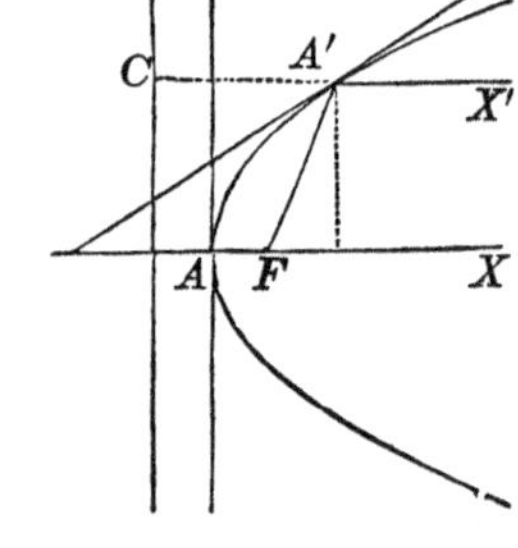


We have, from the last scholium,

$$\frac{2p}{\sin^2 \alpha'} = 2p'.$$

But the equation which determined the direction of the new axis $Y'$, is

$$\text{tang } \alpha' = \frac{p}{b}.$$

But
$$\sin^2 \alpha' = \frac{\text{tang}^2 \alpha'}{1 + \text{tang}^2 \alpha'};$$

hence,
$$\sin^2 \alpha' = \frac{p^2}{b^2 + p^2} = \frac{p}{2a + p}.$$

Substituting this value of $\sin^2\alpha'$ in the first equation, we find

$$2p' = 4a + 2p,$$

or
$$2p' = 4\left(a + \frac{p}{2}\right).$$

But $\left(a + \frac{p}{2}\right)$ is equal to $A'C$, or equal to the distance $A'F$ from the vertex $A'$ to the focus: hence, the parameter of the diameter is four times that distance.

## PROPOSITION VIII. PROBLEM.

*To find the equation of a tangent line to the parabola, when referred to oblique axes.*

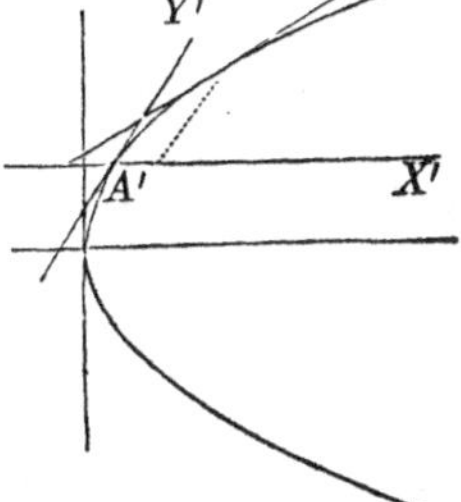


Designate the co-ordinates of the point of tangency by $x''$, $y''$.

The equation of a line passing through this point, will be of the form

$$y - y'' = a(x - x'').$$

Combining this with the equations

$$y^2 = 2p'x,$$

$$y''^2 = 2p'x'',$$

and we find

$$a = \frac{p'}{y''}.$$

Substituting and reducing, we obtain

$$yy'' = p'(x + x''),$$

which is the equation of the tangent.

*Scholium.* If, in the equation of the tangent

$$yy'' = p'(x + x''),$$

we make $y = 0$, we shall have

$$x + x'' = 0,$$

or

$$x = -x'',$$

which shows, that the sub-tangent is bisected at the vertex, and this property may be used in drawing a tangent line to the curve, when it is referred to oblique co-ordinates.

## *Of the Polar Equation of the Parabola and the Measure of its Surface.*

### PROPOSITION IX. PROBLEM.

*To find the general polar equation of the parabola.*

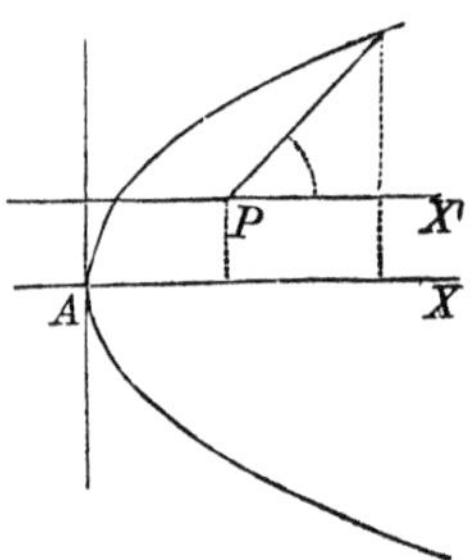


Let $P$ be the pole, and through it draw $PX'$ parallel to the axis $AX$. Then, designating the co-ordinates of the pole by $a$ and $b$, and estimating the angles $v$ from the line $PX'$, we have, for passing from rectangular to polar co-ordinates, the following formulas (Bk. II, Prop. XV):

$$x = a + r \cos v, \qquad y = b + r \sin v.$$

Substituting these values of $x$ and $y$, in the equation

$$y^2 = 2px,$$

and reducing, we obtain

$$r^2 \sin^2 v + 2(b \sin v - p \cos v)r + b^2 - 2pa = 0,$$

which is the general polar equation of the parabola.

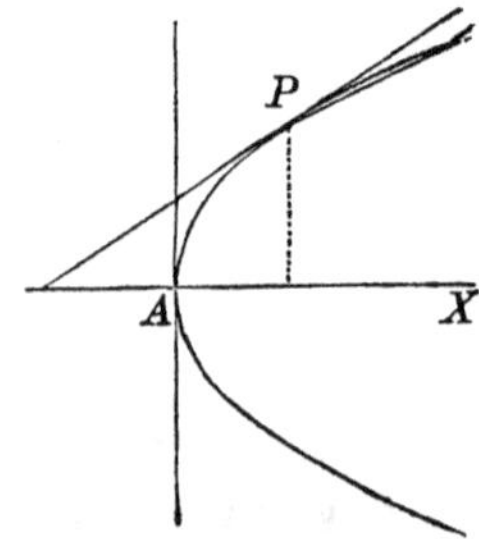


*Scholium* 1. Let us now suppose the pole to be placed on the curve. Its co-ordinates will then satisfy the equation of the curve, and give

$$b^2 - 2pa = 0.$$

The polar equation will then reduce to

$$r^2 \sin^2 v + 2(b \sin v - p \cos v)r = 0,$$

or by dividing by that value of $r$ which is 0, we have,

$$r \sin^2 v + 2(b \sin v - p \cos v) = 0.$$

If now, we make the second value of $r$ equal to 0, the radius-vector will become tangent to the curve, and the equation will give

$$\text{tang } v = \frac{p}{b};$$

the same value for the tangent of the angle formed by the tangent and axis, as was found in Prop. II.

*Scholium* 2. Let the pole be now placed at the focus of the parabola, the co-ordinates of which are

$$a = \frac{p}{2} \quad \text{and} \quad b = 0.$$

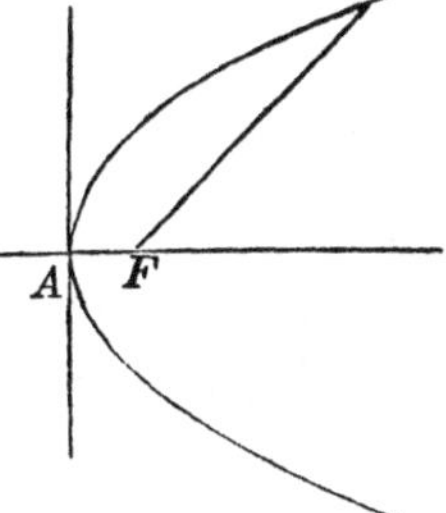


Under this supposition the general polar equation becomes

$$r^2 \sin^2 v - 2p \cos v . r = p^2.$$

Finding the values of $r$ from this equation, we obtain

$$r = \frac{p(\cos v + 1)}{\sin^2 v}, \qquad \text{and} \qquad r = \frac{p(\cos v - 1)}{\sin^2 v}$$

The second value of $r$ is negative for all values of $v$. For, $\cos v$ is less than 1, and the denominator is positive, being a square: hence, this value is to be rejected.

The first value, on the contrary, is positive for all values of $v$, since the numerator and denominator are both positive. By substituting for $\sin^2 v$, $1 - \cos^2 v$, this value may be put under the form

$$r = \frac{p(\cos v + 1)}{(1 + \cos v)(1 - \cos v)},$$

or,
$$r = \frac{p}{1 - \cos v}.$$

In this equation, as well as in the corresponding equation of the ellipse which is expressed under a similar form (Bk. IV, Prop. XII, Sch. 4), the values of $r$ begin at the remote vertex, that is, in the case of the parabola, at an infinite distance from the focus.

If we make $v = 0$: we have

$$r = \frac{p}{0} = \infty.$$

If we make $v = 90^\circ$, we have

$$r = p,$$

that is, half the parameter.

If we make $v = 180^\circ$, we have

$$r = \frac{p}{2} = FA.$$

*Scholium* 3. If it be desirable that the values of $r$ should begin at the nearest vertex, make $v = 180^\circ - v'$, and we shall have

$$\cos v = -\cos v'.$$

Substituting, $-\cos v'$ for $\cos v$, the equation becomes,

$$r = \frac{p}{1 + \cos v'},$$

in which equation the values of $r$ begin at the nearest vertex, and increase from the left to the right, $360^\circ$,

PROPOSITION X. THEOREM.

*The area of any portion of a parabola is equal to two-thirds of the rectangle described on its abscissa and ordinate.*

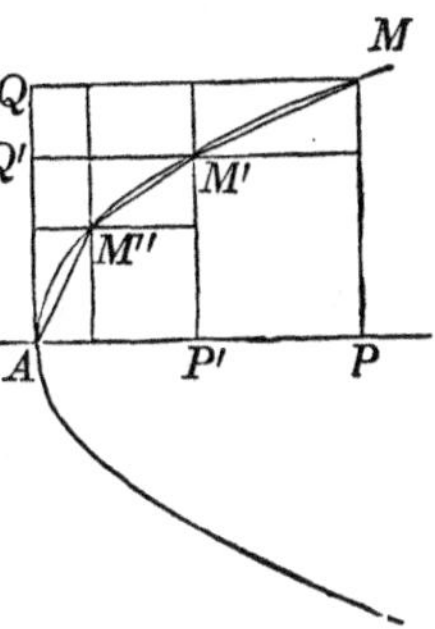


Let $AMP$ be any segment of a parabola. Draw the ordinate $MP$, and draw $MQ$ parallel to $AP$, then will the curvilinear area $AMP$ be two thirds of the rectangle $AQMP$.

Inscribe in the parabola any rectilinear polygon $AM''M' \ldots . M$, and through the vertices $M''$, $M'$, .... $M$, draw lines perpendicular and parallel to $AP$, forming the interior rectangles $M'P$, $M''P'$ ...., and the corresponding exterior rectangles $M'Q$, $M''Q'$ .. . Designate the former by $P$, $P'$ $P''$ ... the latter by $p$, $p'$, $p''$ ...; and the corresponding co-ordinates by $x'$, $y'$, $x''$, $y''$, &c. ——: we shall then have,

$$P = y''(x' - x'') \qquad p = x''(y' - y''),$$

which gives,

$$\frac{P}{p} = \frac{y''(x' - x'')}{x''(y' - y'')}.$$

But, since the points, $M$, $M'$, $M''$, ... are on the curve,

$$y'^2 = 2px', \qquad y''^2 = 2px'',$$

which gives,

$$x' - x'' = \frac{y'^2 - y''^2}{2p}, \quad \text{and} \quad x'' = \frac{y''^2}{2p}$$

Substituting these values, and we obtain

$$\frac{P}{p} = \frac{y''(y'^2 - y''^2)}{y''^2(y' - y'')} = \frac{y' + y''}{y''}.$$

Applying similar reasoning to each of the rectangles, we shall have,

$$\frac{P}{p}=\frac{y'+y''}{y''}.$$

$$\frac{P'}{p'}=\frac{y''+y'''}{y'''}.$$

$$\frac{P''}{p''}=\frac{y'''+y''''}{y''''} \text{ \&c.}$$

The inscribed polygon being entirely arbitrary, we can place the vertices of the angles in such a manner that the ordinates passing through them shall be in geometrical progression. We shall then have,

$$y' \; : \; y'' \; :: \; y'' \qquad y''',$$

or, $$y'-y'' \; : \; y'' \; :: \; y''-y''' \; : \; y'''$$

(Geom. Bk. II, Prop. VI);

hence, $$\frac{y'-y''}{y''}=\frac{y''-y'''}{y'''}.$$

Therefore, the difference between two successive ordinates divided by the less, is constant. If we designate this ratio by $q$, we shall have

$$y'-y''=qy'' \qquad \text{or} \qquad y'+y''=y''(2+q),$$

$$v''-y'''=qy''' \qquad\qquad y''+y'''=y'''(2+q),$$

$$y'''-y''''=qy'''' \qquad\qquad y'''+y''''=y''''(2+q).$$

Substituting these values, the ratios of the rectangles

become, $$\frac{P}{p}=2+q,$$

$$\frac{P'}{p'}=2+q,$$

$$\frac{P''}{p''}=2+q,$$

$$\frac{P'''}{p'''}=2+q.$$

But the sum of all the consequents divided by the sum of the antecedents will give the same ratio as either consequent divided by its antecedent (Geom. Bk. II, Prop. X): hence,

$$\frac{P + P' + P'' + P''' \quad \ldots\ldots \&c.}{p + p' + p'' + p''' \quad \ldots\ldots \&c.} = 2 + q.$$

In this expression the numerator is equal to the sum of the interior rectangles, and the denominator to the sum of the exterior ones.

Since $q$ is the ratio of the difference between two successive ordinates divided by the less, if $q$ be diminished, the sum of the interior rectangles will approach to the curvilinear area $AMP$, and the sum of the exterior rectangles to the area $AQM$. If then, we pass to the limit by making $q = 0$, the numerator will become the curvilinear area $AMP$, and the denominator, the area $AQM$. Designating the former by $S$, and the latter by $s$, we have,

$$\frac{S}{s} = 2,$$

which gives, $$\frac{S + s}{s} = 3:$$

hence, $$S = \frac{2}{3}(S + s).$$

But $S + s$ is equal to the area of the rectangle $APMQ$, therefore the area of the parabola is two-thirds of that area.

*Corollary.* The exterior portion $AQM$ is one third the area of the rectangle.

# BOOK VI.

## *Of the Hyperbola.*

An *hyperbola* is a curve in which the difference of two straight lines, drawn from any one of its points to two fixed points, is constantly equal to a given line.

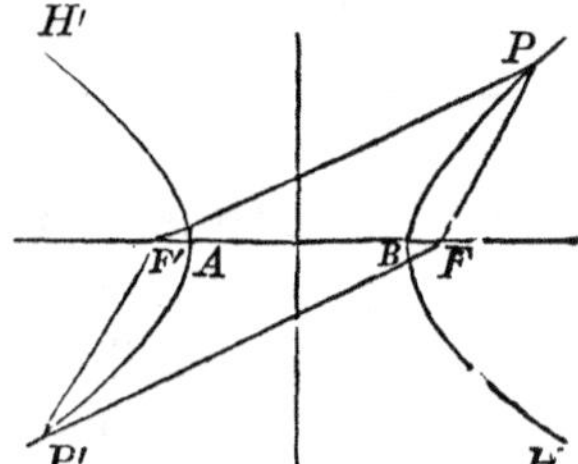


Thus, if $F$ and $F'$ be the two fixed points, $AB$ the given line, and $F'P-FP$ constantly equal to $AB$, for every position of the point $P$, $PBH$ will be a portion of the hyperbola.

If $FP$ is greater than $F'P$, let it be represented by $FP'$; and then, if $FP'-F'P'$ is constantly equal to the given line $AB$, the point $P'$ will describe the remaining portion of the hyperbola $P'AH'$.

The two curves, $P'AH'$, $HBP$, are called branches of the hyperbola.

The fixed points $F$ and $F'$, are called the *foci* of the hyperbola. Hence, each branch has two foci, one lying within the curve, and one without it.

The definition of an hyperbola affords an easy method of describing it mechanically.

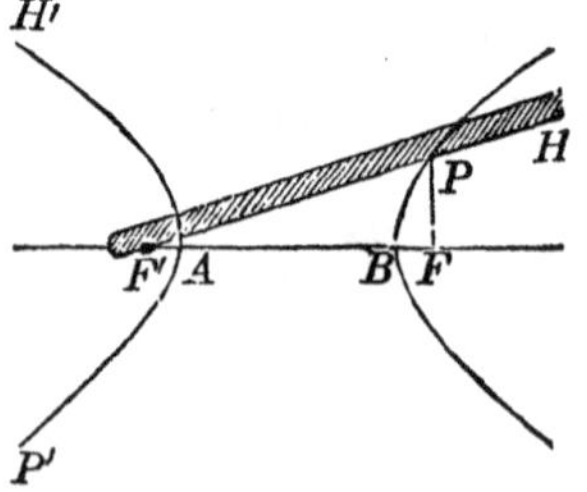


Take a ruler, longer than the distance $F'F$, and fasten one of its extremities at the focus $F'$. At the other extremity $H$, attach a thread, of such a length, that the length of the ruler shall exceed the length of the thread by the given line $AB$. Attach the other extremity of the thread to the focus $F$.

Place a pencil against the thread, and press it against the ruler, keeping the thread constantly stretched while the ruler is turned around $F'$ as a centre. The point of the pencil will describe a branch of an hyperbola.

For, $PF+PH$ is equal to the length of the thread, to which if we add $AB$, we shall have the length of the ruler Hence,

$$F'P+PH=FP+PH+AB,$$

or $$F'P-FP=AB;$$

therefore, $P$ is a point of the hyperbola.

If one extremity of the ruler be attached to the focus $F$, the branch $P'AH'$ may be described.

### PROPOSITION I. PROBLEM.

*To find the equation of an hyperbola.*

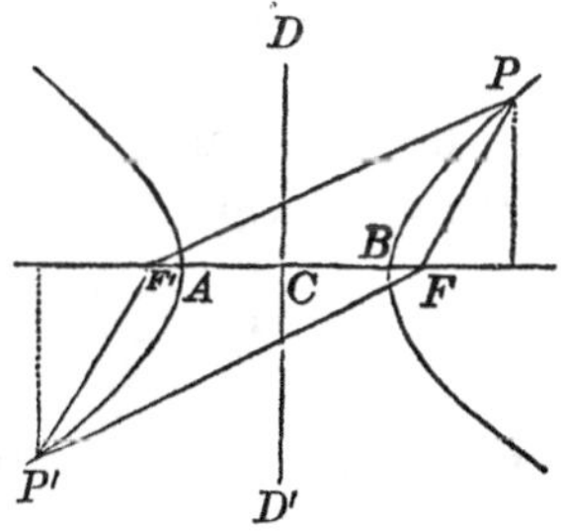


Let $F$ and $F'$ be the foci, and denote the distance between them by $2c$. Let $P$ be any point of the curve, and designate the distance $FP$ by $r$, and $F'P$ by $r'$; and let $2A$ represent the given line, to which the difference $F'P-PF$ is to be equal.

Through $C$, the middle point of $F'F$, draw $CD$ perpendicular to $F'F$, and let $C$ be the origin of a system of rectangular co-ordinates, of which $AB$, $DD'$ are the axes. Let $x$ and $y$ represent the co-ordinates of the point $P$.

The square of the distance between any two points of

which the co-ordinates are $x$, $y$, and $x'$, $y'$, is (Bk. II, Prop. III),

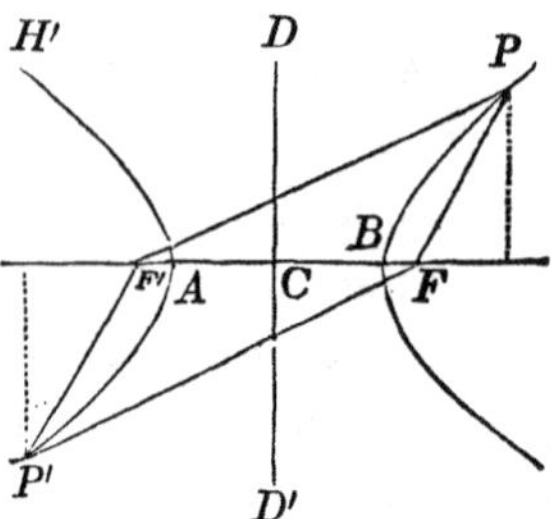


$$(y-y')^2+(x-x')^2.$$

If this line passes through the point $F$, of which the co-ordinates are $y'=0$, $x'=c$, we shall have

$$\overline{FP}^2=r^2=y^2+(x-c)^2;$$

and if we pass the line through the point $F'$, of which the co-ordinates are $y'=0$ and $x'=-c$, we shall have

$$\overline{F'P}^2=r'^2=y^2+(x+c)^2.$$

If we add and subtract these two equations, we obtain

$$r^2+r'^2=2(y^2+x^2+c^2), \quad \text{and} \quad r'^2-r^2=4cx.$$

But we have, from the property of the hyperbola,

$$r'-r=2A.$$

Combining this with the last equation, we obtain

$$r'=A+\frac{cx}{A}, \quad \text{and} \quad r=-A+\frac{cx}{A}.$$

Squaring these values and substituting in the equation, of which the first member is $r^2+r'^2$, and there will result,

$$A^2+\frac{c^2x^2}{A^2}=y^2+x^2+c^2,$$

or $$A^2(y^2+x^2)-c^2x^2=A^2(A^2-c^2),$$

which is the equation of the hyperbola referred to the axes $CB$, $CD$. If we suppose $r>r'$, we shall obtain an equation of the same form for the branch $P'AH'$.

The equation of the hyperbola takes the most convenient form when it is expressed in terms of the co-ordinates of its points, and the distances which the curve cuts off from the co-ordinate axes.

To place the equation under this form, let us make $x = 0$, this will give

$$y^2 = A^2 - c^2,$$

or

$$y = \pm\sqrt{A^2 - c^2};$$

and since $c$ is greater than $A$ the quantity under the radical is essentially negative, and therefore, the value of $y$ is imaginary.

This result shows that conditions have been introduced into the equation of the curve, which are incompatible with each other (Alg. Art. 147). The incompatible condition is, that which attributes to $x$ the value 0; for this condition requires that the curve should have one or more of its points on the axis of $Y$, while the law by which the curve was described, did not permit it to fulfil that condition.

The imaginary value of $y$, for $x = 0$, is, however, a constant quantity, and may be introduced into the equation of the curve.

Let us make

$$y = \pm\sqrt{A^2 - c^2} = \pm B\sqrt{-1} = CD \quad \text{or} \quad CD',$$

we shall then have

$$A^2 - c^2 = -B^2, \quad \text{and} \quad c^2 = A^2 + B^2.$$

Substituting these values of $c^2$ in the equation of the curve, and reducing, we obtain

$$A^2y^2 - B^2x^2 = -A^2B^2;$$

in which, if we make $y = 0$, we shall have

$$x = \pm A = CB \quad \text{or} \quad CA$$

*Scholium* 1 The point $C$ which is equidistant from $F$ and $F'$, is called the *centre* of the hyperbola.

Every straight line passing through the centre and terminating in the curve, is called a *diameter.*

The diameter $AB$, which passes through the foci, is called the *transverse axis:* and since $2CA$, or $AB$ is equal to $2A$, it follows, that *the difference of the two lines drawn from any point of the curve to the two foci is equal to the transverse axis.*

The line $DD'$, which is perpendicular to the transverse axis at the middle point, and equal to $2B$, is called the *conjugate axis.*

In the equation of the hyperbola,

$$A^2y^2 - B^2x^2 = - A^2B^2,$$

$A$ and $B$ represent the semi-axes, and $x$ and $y$ the general co-ordinates of the curve. It is called, *the equation of the hyperbola referred to its centre and axes.*

In comparing this equation with the equation of the ellipse, it is seen that the two are similar in every respect, excepting in the sign of $B^2$, which is minus in the hyperbola and plus in the ellipse. We may, therefore, pass from one equation to the other, by simply substituting for $B$, $B\sqrt{-1}$.

*Scholium* 2. If in the equation of the hyperbola,

$$A^2y^2 - B^2x^2 = - A^2B^2,$$

we change $y$ into $x$, and $x$ into $y$, it becomes,

$$A^2x^2 - B^2y^2 = - A^2B^2,$$

or,

$$B^2y^2 - A^2x^2 = A^2B^2.$$

If in this equation we suppose $x = 0$, the corresponding value of $y$ will be real, and if we make $y = 0$, the corres ponding value of $x$ will be imaginary. The transverse axis of the curve will then coincide with the axis of $Y$, but its value will be the same as before, viz. $2A$.

If on the conjugate axis $DD'$ an hyperbola be described. corresponding to the equation

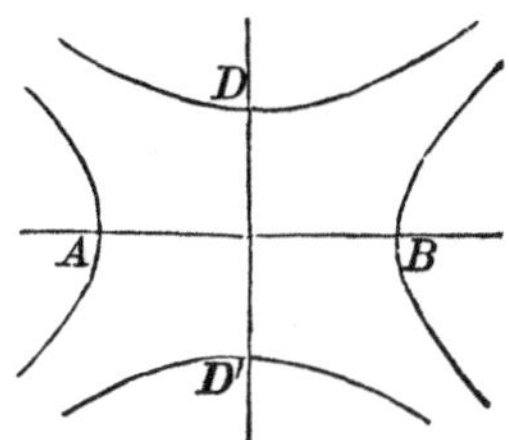


$$B^2x^2 - A^2y^2 = -A^2B^2,$$

it is said to be conjugate to the hyperbola described on the transverse axis $AB$. The transverse axis of one hyperbola is the conjugate axis of the other, and reciprocally.

*Scholium* 3. If through the centre of the hyperbola any line be drawn, its equation will be of the form

$$y = ax.$$

If we combine this equation with the equation of the hyperbola

$$A^2y^2 - B^2x^2 = -A^2B^2,$$

we shall obtain the co-ordinates of the points $H$ and $H'$, in which the diameter intersects the hyperbola.

If we designate the co-ordinates of $H$ by $x'$, $y'$, and the co-ordinates of $H'$ by $x''$, $y''$, we shall find, after eliminating

$$x' = AB\sqrt{\frac{-1}{A^2a^2 - B^2}}, \qquad y' = ABa\sqrt{\frac{-1}{A^2a^2 - B^2}},$$

$$x'' = -AB\sqrt{\frac{-1}{A^2a^2 - B^2}}, \qquad y'' = -ABa\sqrt{\frac{-1}{A^2a^2 - B^2}}.$$

But since the co-ordinates of $H'$ are the same as those of $H$, excepting that the signs are both negative, it follows that $CH = CH'$, that is, *every diameter of an hyperbola is bisected at the centre.*

*Scholium* 4. If $B$ be made equal to $A$ the equation of the hyperbola will become,

$$y^2 - x^2 = -A^2;$$

the hyperbola is then said to be *equilateral*, which corresponds to the case in which the ellipse becomes a circle.

*Scholium* 5. By a course of reasoning similar to that pursued in Bk. IV, Prop. I, Sch. 5, we find the following analytical conditions for determining the position of a point with respect to the hyperbola.

| | |
|---|---|
| Without the hyperbola | $A^2y^2 - B^2x^2 + A^2B^2 > 0,$ |
| on the hyperbola | $A^2y^2 - B^2x^2 + A^2B^2 = 0,$ |
| within the hyperbola | $A^2y^2 - B^2x^2 + A^2B^2 < 0.$ |

*Scholium* 6. If we place the equation of the hyperbola under the form

$$y = \pm \frac{B}{A}\sqrt{x^2 - A^2},$$

we see, that every value of $x$, either plus or minus, which is less than $A$, will render $y$ imaginary.

If we make $x = \pm A$, $y$ will become equal to 0, which shows that the curve cuts the axis of $X$ at two points, one on the positive and the other on the negative side of abscissas, and each at a distance from the centre equal to $A$.

Every value of $x$, either plus or minus, which is greater than $A$, will give two equal values of $y$ with contrary signs; hence, the curve extends itself indefinitely in the direction of $x$ positive and $x$ negative, and is symmetrical with respect to the transverse axis.

*Scholium* 7. If we transfer the origin of co-ordinates from the centre $C$ to $A$, one extremity of the transverse axis, the equations of transformation (Bk. II, Prop. X) will reduce to

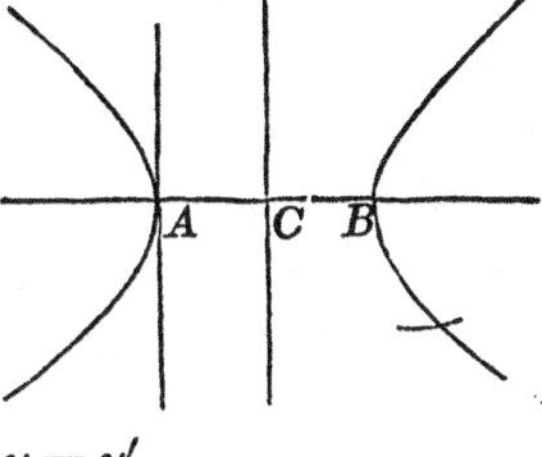


$$x = -A + x', \qquad y = y'.$$

Substituting these values in the equation of the hyperbola, it reduces to

$$A^2y'^2 - B^2x'^2 + 2B^2Ax' = 0;$$

which may be put under the form

$$y'^2 = \frac{B^2}{A^2}(x'^2 - 2Ax'),$$

which is the equation of the hyperbola referred to the vertex $A$ as an origin of co-ordinates.

If we refer it to the vertex B, as an origin, the equation will become

$$y'^2 = \frac{B^2}{A^2}(2Ax' + x'^2).$$

*Scholium* 8. The property, that the difference of the two lines drawn from any point of the curve to the foci, is equal to the transverse axis, affords an easy method of describing the hyperbola by points, when the transverse axis and the foci are known.

Let $AB$ be the transverse axis of an hyperbola, and $F$ and $F'$ the foci.

From the focus $F'$ lay off a distance $F'N$ equal to the transverse axis, and take any other distance, as $F'\,H$, greater than $F'B$.

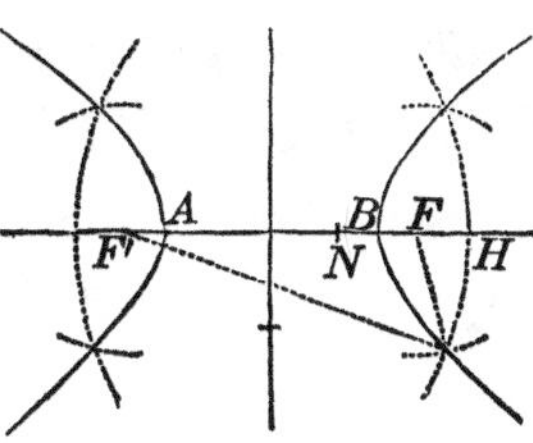

With $F'$ as a centre, and $F'H$ as a radius, describe the arc of a circle. Then, with $F$ as a centre, and $HN$ as a radius, describe an arc intersecting the arc before described at $p$ and $q$, and they will be points of the hyperbola.

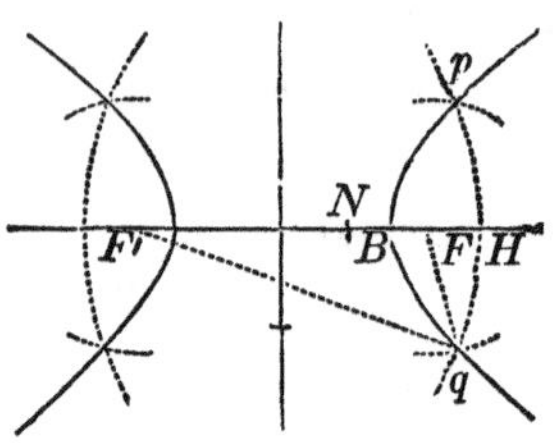


If with $F$ as a centre, and $F'H$ as a radius, an arc be described, and a second arc be described with $F'$ as a centre and $HN$ as a radius, two points in the other branch of the curve will be determined.

*Scholium* 9. The property, that the square of the distance from the centre to either focus is equal to the sum of the squares of the semi-axes, affords an easy construction for the foci when the axes are known.

For, from the vertex $B$, draw $BH$ perpendicular to $AB$, and make it equal to the semi-conjugate axis. Join $H$ and the centre $C$. Then, with $C$ as a centre and $CH$ as a radius, describe a semi-circumference, intersecting $AB$ produced in $F$ and $F'$, and these points will be the foci.

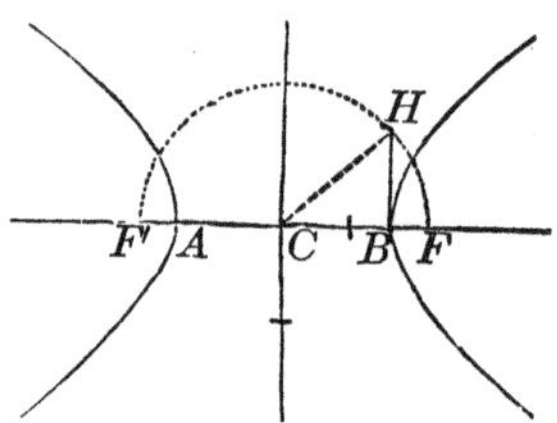


PROPOSITION II. THEOREM.

*The squares of the ordinates are to each other, as the rectangles of the segments from the foot of each ordinate respectively, to the vertices of the transverse axis.*

The equation of the hyperbola $HBP$ referred to the vertex $B$ (Prop. I, Sch. 7) is,

$$y^2 = \frac{B^2}{A^2}(2Ax + x^2).$$

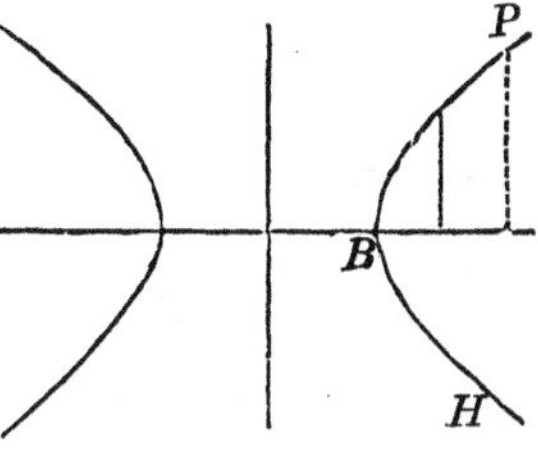


If we designate a particular ordinate by $y'$ and its abscissa by $x'$, and a second ordinate by $y''$ and its abscissa by $x''$, we shall have,

$$y'^2 = \frac{B^2}{A^2}(2Ax' + x'^2), \quad \text{and} \quad y''^2 = \frac{B^2}{A^2}(2Ax'' + x''^2).$$

Dividing one equation by the other, we obtain,

$$\frac{y'^2}{y''^2} = \frac{(2A + x')x'}{(2A + x'')x''}, \qquad \text{or,}$$

$$y'^2 \ : \ y''^2 \ :: \ (2A + x')x' \ : \ (2A + x'')x''.$$

in which it is evident that the segments, are

$$2A + x',\ x', \qquad \text{and} \qquad 2A + x'',\ x''.$$

## PROPOSITION III. THEOREM.

*If through the vertices of the transverse axis two supplementary chords be drawn, the product of the tangents of the angles which they form with it, on the same side, will be equal to the square of the ratio of the semi-axes.*

The equation of the chords will be of the form,

$$y = a'(x + A),$$

$$y = a\,(x - A).$$

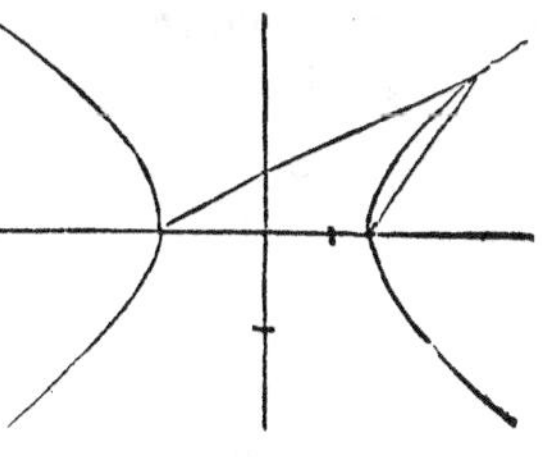

Combining them, we obtain,

$$y^2 = aa'(x^2 - A^2).$$

Combining this with the equation of the hyperbola

$$y^2 = \frac{B^2}{A^2}(x^2 - A^2),$$

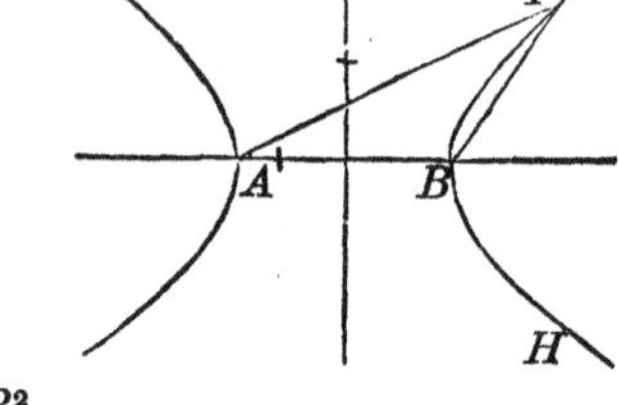


and we find

$$aa' = \frac{B^2}{A^2}.$$

*Scholium* 1. In the equation

$$aa' = \frac{B^2}{A^2},$$

there are two undetermined quantities $a$ and $a'$: hence, an infinite number of supplementary chords may be drawn through the extremities of the diameter $AB$.

If, however, a value be assigned to $a$ or $a'$, that is, if one of the supplementary chords be given in position, the equation of condition will determine the direction of the other, and therefore, the corresponding supplementary chord will also be known.

*Scholium* 2. If the chords are drawn to a point $P$ in the branch $HBP$, the tangents $a$ and $a'$ will be both positive; if drawn to a point in the other branch, they will both be negative.

*Scholium* 3. If the hyperbola is equilateral, $A = B$, and there will result.

$$aa' = 1,$$

which shows, that *the sum of the two acute angles formed by the supplementary chords with the transverse axis, on the same side, is equal to* 90°.

### PROPOSITION IV. PROBLEM.

*To find the equation of a straight line which shall be tangent to an hyperbola.*

The equation of the hyperbola is

$$A^2y^2 - B^2x^2 = -A^2B^2.$$

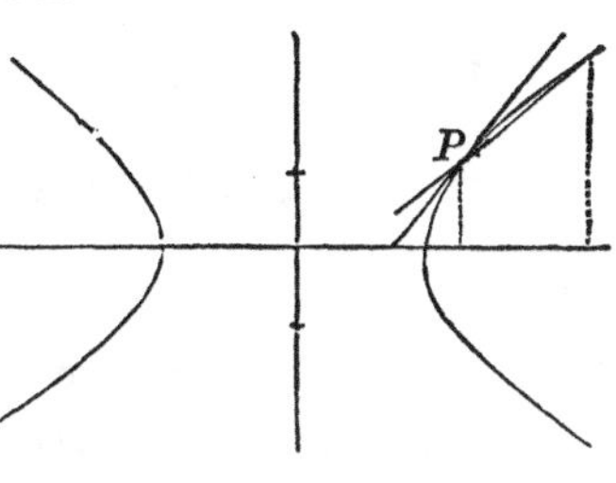


Take any point of the curve, as $P$, and designate its co-ordinates by $x''$, $y''$, and assume also a second point of the curve, and represent its co-ordinates by $x'$, $y'$.

The equation of a straight line passing through these two points (Bk. II, Prop. V), is

$$y - y'' = \frac{y'' - y'}{x'' - x'}(x - x'').$$

But since the two assumed points belong to the hyperbola, we have the equations

$$A^2y''^2 - B^2x''^2 = -A^2B^2,$$

$$A^2y'^2 - B^2x'^2 = -A^2B^2.$$

By subtracting the second equation from the first, we obtain

$$A^2(y''^2 - y'^2) - B^2(x''^2 - x'^2) = 0,$$

or $$A^2(y'' + y')(y'' - y') - B^2(x'' + x')(x'' - x') = 0;$$

whence, $$\frac{y'' - y'}{x'' - x'} = \frac{B^2(x'' + x')}{A^2(y'' + y')}.$$

Substituting this value in the equation of the secant line, it becomes

$$y - y'' = \frac{B^2}{A^2}\frac{(x'' + x')}{(y'' + y')}(x - x'').$$

If we now suppose the co-ordinates of the two points to become equal, the points will unite, and the secant line will become tangent to the curve. This supposition will reduce the equation to

$$y - y'' = \frac{B^2x''}{A^2y''}(x - x''), \qquad a = \frac{B^2x''}{A^2y''},$$

or

$$A^2yy'' - B^2xx'' = -A^2B^2,$$

which is the equation of a tangent line to the hyperbola

*Scholium* 1. This method of determining the equation of a tangent, may be employed in finding the equation of a tangent to the circle, the ellipse, or the parabola.

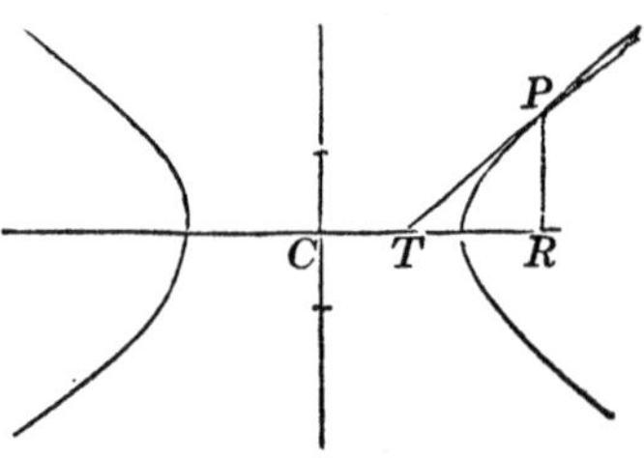


*Scholium* 2. If in the equation of the tangent

$$A^2yy'' - B^2xx'' = -A^2B^2,$$

we make $y = 0$, we shall nave,

$$CT = x = \frac{A^2}{x''},$$

Subtracting this from $CR = x''$, and we obtain,

$$TR = \frac{x''^2 - A^2}{x''},$$

which is the value of the subtangent.

### PROPOSITION V. PROBLEM.

*To find the equation of a normal line to the hyperbola.*

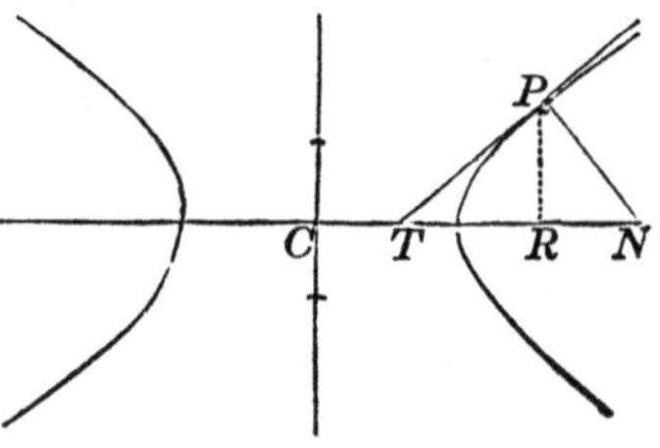


Since the normal passes through the point of tangency its equation will be of the form,

$$y - y'' = a'(x - x''),$$

and since it is perpendicular

to the tangent, we shall have,

$$aa' + 1 = 0.$$

But we have already found,

$$a = \frac{B^2}{A^2}\frac{x''}{y''},$$

hence,
$$a' = -\frac{A^2}{B^2}\frac{y''}{x''}.$$

Substituting this value, and the equation of the normal will become.

$$y - y'' = -\frac{A^2}{B^2}\frac{y''}{x''}(x - x'').$$

*Scholium.* To find the point in which the normal intersects the axis of $X$ make $y = 0$, and we have,

$$CN = x = \frac{A^2 + B^2}{A^2}x'',$$

and by subtracting $x''$, we find the subnormal

$$RN = \frac{B^2x''}{A^2}.$$

## PROPOSITION VI. THEOREM.

*If one of the supplementary chords of an hyperbola is parallel to a tangent line to the curve, the other will be parallel to the diameter which passes through the point of contact:* and conversely,

*If one of the chords be parallel to the diameter which passes through the point of contact, the other will be parallel to the tangent line*

The equation of a line passing through the centre of the hyperbola, is of the form

$$y = a'x.$$

The condition of its passing through the point of contact, will give

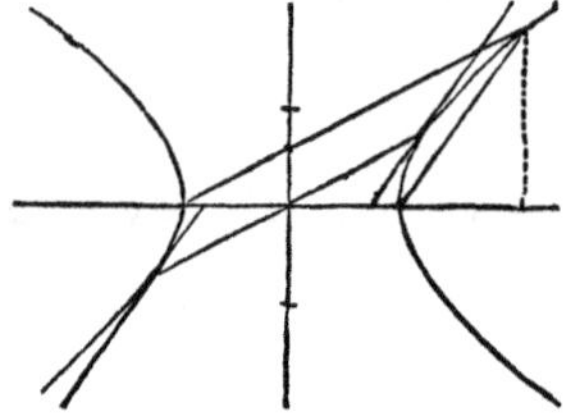

$$y'' = a'x'',$$

whence, $$a' = \frac{y''}{x''}.$$

But we have found for the tangent of the angle, which the tangent line makes with the transverse axis,

$$a = \frac{B^2x''}{A^2y''}.$$

Multiplying the members of these equations together we obtain

$$aa' = \frac{B^2}{A^2}$$

By comparing this equation with the equation in Prop. III, we see, that the product of the tangents of the angles which the diameter and tangent make with the transverse axis, is equal to the product of the tangents of the angles which the supplementary chords form with the axis. Hence, if in these equations we make

$$a = a,$$

we shall have, $$a' = a';$$

that is, if one of the chords is parallel to the tangent, the other will be parallel to the diameter passing through the point of contact. Or, if we make

$$a' = a',$$

we shall have, $$a = a;$$

that is, if one of the chords be made parallel to the diameter the other will be parallel to the tangent.

*Scholium.* These properties afford an easy method of drawing a tangent line to an hyperbola at a given point of the curve.

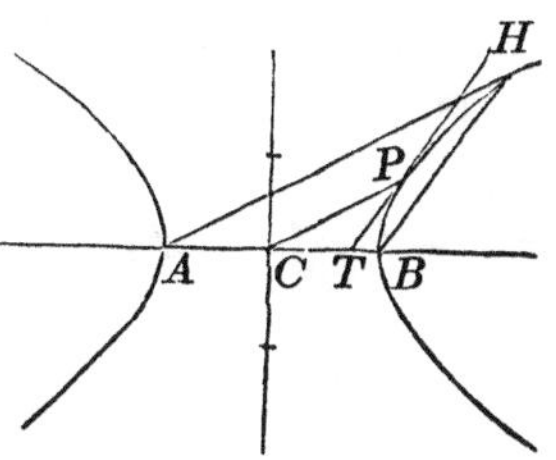


Let $C$ be the centre of the hyperbola, $AB$ its transverse axis, and $P$ the given point of the curve at which the tangent is to be drawn.

Through $P$ draw the semi-diameter $PC$, and through $A$ draw the supplementary chord $AH$ parallel to it. Then draw the other supplementary chord $BH$, and through $P$, draw $PT$ parallel to $BH$; then will $PT$ be the tangent required.

In a similar manner, a line might be drawn tangent to the curve, and parallel to a given line.

## PROPOSITION VII. THEOREM.

*If a line be drawn tangent to an hyperbola at any point, and two lines be drawn from the same point to the foci, the lines drawn to the foci will make equal angles with the tangent.*

Let $C$ be the centre of the hyperbola, $PT$ the tangent line, and $PF$, $PF'$, the two lines drawn to the foci.

Denote the distance $CF = \sqrt{A^2 + B^2}$ by $c$, $CF'$ by $-c$, the angle $FPT$ by $v$, and the tangents of the angles $PFX$, $PTX$, by $a'$ and $a$. We shall then have

$$\text{tang } v = \frac{a' - a}{1 + a'a}.$$

But the equation of $FP$ passing through two points, of which the co-ordinates are $x' = c$, $y' = 0$, and $x''$ $y''$, is

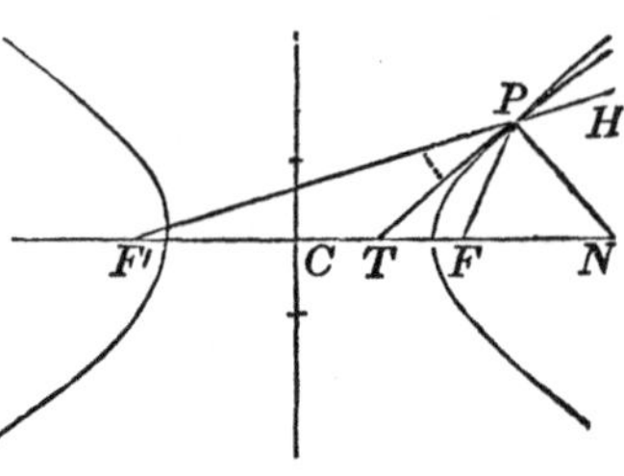


$$y = \frac{-y''}{c - x''}(x - x');$$

hence,
$$a' = \frac{-y''}{c - x''}.$$

We also have
$$a = \frac{B^2 x''}{A^2 y''}.$$

Substituting these values, and recollecting that,

$$A^2 y''^2 - B^2 x''^2 = -A^2 B^2,$$

we find
$$\text{tang } v = \frac{B^2}{cy''}.$$

If we designate the angle $F'PT$ by $v'$, and the tangent of $PF'C$ by $a'$, we shall have

$$\text{tang } v' = \frac{a - a'}{1 + aa'},$$

which reduces to
$$\frac{B^2}{cy''};$$

therefore, the tangent line bisects the angle $F'PF$.

*Corollary.* The normal line $PN$, bisects the outward angle $FPH$, formed by the two lines drawn to the foci.

*Scholium* 1. The relation between the angles formed by the tangent line, and the lines drawn to the foci, enables us to draw a tangent to the curve at a given point.

Let $P$ be the given point. From $P$ draw $PF$, $PF'$ to the foci. Lay off on $PF'$, $PG = PF$, and draw $FG$. From $P$, draw $PT$ perpendicular to $FG$, and it will be tangent to the hyperbola at $P$, since it bisects the angle $F'PF$.

*Scholium* 2. The same properties also enable us to draw a tangent to the hyperbola through a given point without the curve.

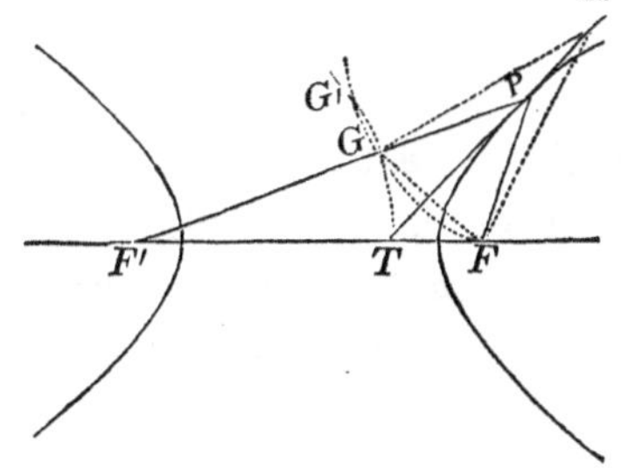


Let $H$ be the given point. With this point as a centre, and $HF$ as a radius, describe the arc of a circle. With $F'$ as a centre, and a radius equal to the transverse axis, describe the arc of a circle intersecting the former at $G$ and $G'$. Draw $F'G$, cutting the curve in $P$ Through $P$ draw $HPT$, and it will be tangent to the hyper bola at $P$.

For, if we draw $HF$, $HG$, we shall have $HF = HG$ by construction; and since $P$ is a point of the hyperbola, and $F'G$ equal to the transverse axis, we shall have $PF = PG$: hence, $PT$ is perpendicular to $FG$; and since the triangle $FGP$ is isosceles, $PT$ will bisect the angle $F'PF$, and will therefore be tangent to the hyperbola.

*Scholium* 3. The two arcs described with the centres $F'$ and $H$, intersect each other in two points, $G$ and $G'$; a line may, therefore, be drawn through $F'$, and either of these points, thus giving two points of tangency.

It may also be shown analytically, that two tangent lines can be drawn to the hyperbola from a given point without the curve.

For, if the tangent pass through a given point, of which

the co-ordinates are $x'$, $y'$, its equation will become

$$A^2y'y''-B^2x'x''=-A^2B^2,$$

and the equation of the hyperbola for the point of tangency becomes,

$$A^2y''^2-B^2x''^2=-A^2B^2.$$

In these two equations all the quantities are known, except $x''$, $y''$, which may therefore be found; and since the equation of the tangent is of the first degree, with respect to $x''$ and $y''$, the equation which results from their combination will be of the second degree, and will therefore give two values for $x''$ and two values for $y''$, which values will be real if the given point lies without the curve.

## *Of the Hyperbola referred to its Conjugate Diameters.*

1. Two diameters of an hyperbola are said to be conjugate to each other, when either of them is parallel to the two tangent lines which may be drawn through the vertices of the other.

2. Since two supplementary chords may always be drawn respectively parallel to a diameter and the tangent lines through its vertices, it follows, that two supplementary chords may always be drawn, respectively parallel to two conjugate diameters.

If, therefore, we designate the tangents of the angles which two conjugate diameters make with the transverse axis, by $a$ and $a'$, these tangents must satisfy the equation

$$aa'=\frac{B^2}{A^2}.$$

Let us designate the corresponding angles by $\alpha$ and $\alpha'$

We shall then have

$$a = \frac{\sin \alpha}{\cos \alpha}, \qquad a' = \frac{\sin \alpha'}{\cos \alpha'}.$$

Substituting these values in the last equation, and reducing, we obtain

$$A^2 \sin \alpha \sin \alpha' - B^2 \cos \alpha \cos \alpha' = 0,$$

which expresses the relation between the angles which two conjugate diameters form with the transverse axis.

3 It may be easily shown here, as in the ellipse, that the axes of the curve are conjugate diameters ; and also, that they are the only conjugate diameters which are at right angles to each other

PROPOSITION VIII. PROBLEM.

*To find the equation of the hyperbola referred to its centre and conjugate diameters.*

The equation of the hyperbola referred its centre and axes, is

$$A^2y^2 - B^2x^2 = -A^2B^2.$$

The formulas for passing from rectangular to oblique co-ordinates, the origin remaining the same, are

$$x = x' \cos \alpha + y' \cos \alpha', \qquad y = x' \sin \alpha + y' \sin \alpha'.$$

Squaring these values of $x$ and $y$, and substituting in the equation of the hyperbola, we have.

$$\left.\begin{array}{l}(A^2 \sin^2\alpha' - B^2 \cos^2\alpha')y'^2 + (A^2 \sin^2\alpha - B^2 \cos^2\alpha)x'^2 \\ \quad + 2(A^2 \sin \alpha \sin \alpha' - B^2 \cos \alpha \cos \alpha')x'y'\end{array}\right\} = -A^2B^2$$

But the condition that the new axes shall be conjugate diameters, gives

$$A^2 \sin \alpha \sin \alpha' - B^2 \cos \alpha \cos \alpha' = 0;$$

hence, the equation reduces to

$$(A^2 \sin^2\alpha' - B^2 \cos^2\alpha')y'^2 + (A^2 \sin^2\alpha - B^2\cos^2\alpha)x'^2 = -A^2B^2.$$

If we suppose in succession, $y'=0$, $x'=0$, and represent by $A'$ and $B'$ the corresponding abscissa and ordinate, we find

$$A'^2 = \frac{-A^2B^2}{A^2 \sin^2\alpha - B^2 \cos^2\alpha}, \quad B'^2 = \frac{-A^2B^2}{A^2 \sin^2\alpha' - B^2 \cos^2\alpha'}.$$

If we suppose the semi-diameter $A'$ to be real, we shall have

$$A^2 \sin^2\alpha < B^2 \cos^2\alpha, \quad \text{or} \quad \text{tang } \alpha < \frac{B}{A}.$$

But, $\quad \text{tang } \alpha \text{ tang } \alpha' = \frac{B^2}{A^2}; \quad$ hence,

$$\text{tang } \alpha' > \frac{B}{A}, \quad \text{or} \quad A^2 \sin^2\alpha' > B^2 \cos^2\alpha';$$

hence, $B'^2$ will be negative.

The supposition, therefore, which renders $A'^2$ positive, or $A'$ real, gives $B'^2$ negative, or $B'$ imaginary. Attributing to $B'^2$ its proper sign, we have

$$A'^2 = \frac{-A^2B^2}{A^2 \sin^2\alpha - B^2 \cos^2\alpha}, \quad -B'^2 = \frac{-A^2B^2}{A^2 \sin^2\alpha' - B^2 \cos^2\alpha'}.$$

Finding the values of the denominators in these equations, and substituting these values in the general equation, and reducing, we obtain

$$A'^2y'^2 - B'^2x'^2 = -A'^2B'^2,$$

or, omitting the accents of $x'$ and $y'$, since they are general variables, we obtain

$$A'^2y^2 - B'^2x^2 = -A'^2B'^2,$$

for the equation of the hyperbola, referred to its centre and conjugate diameters.

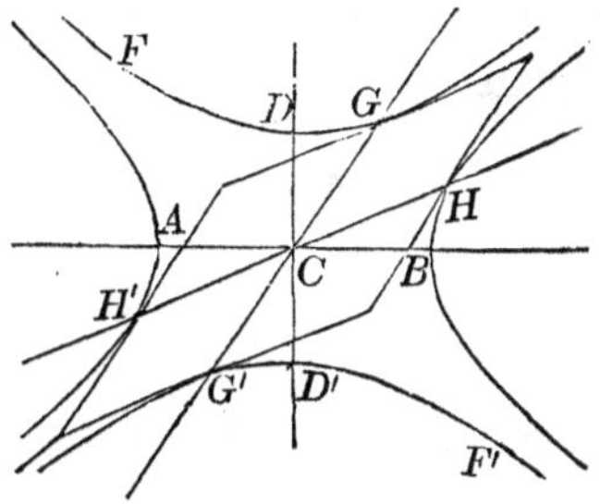


We have already seen (Prop. I), that when the transverse axis $AB$ is real, the conjugate axis $DD'$ will be imaginary, and reciprocally: that is, the two axes will not intersect the same hyperbola. The last proposition proves the same property for any two conjugate diameters.

If then, $2A'$ designates the diameter $H'H$, $2B'$ will designate the diameter $G'G$ terminating in the conjugate hyperbola; and each will be parallel to the two tangent lines drawn through the vertices of the other.

If $B'$ were made real, $A'$ would be imaginary, and the equation would represent the curves $FDG$, $F'D'G'$.

*Scholium* 1. The equation of the hyperbola, referred to its centre and conjugate diameters, being of the same form as when referred to its centre and axes, it follows that every value of $x$ will give two values of $y$ with contrary signs; or if $B'$ were real, every value of $y$ would give two equal values of $x$ with contrary signs: hence, each hyperbola is symmetrical with respect to the diameter which it intersects: that is, *either diameter will bisect all chords drawn parallel to the other and terminated by the curve.*

*Scholium* 2. If the curve of an hyperbola be traced on a plane the centre and axes are found in a manner entirely similar to that pursued in (Bk. IV, Prop. IX, Sch. 2).

*Scholium* 3. It may also be readily shown that the squares of the ordinates to either diameter, are proportional to the rectangles of the corresponding segments from the foot of the ordinates respectively, to the vertices of the diameter.

*Scholium* 4. The *parameter* of any diameter is a third

proportional to the diameter and its conjugate Thus, if $P$ designate the parameter of the diameter $2A'$, we shall have,

$$2A' \ : \ 2B' \ :: \ 2B' \ : \ P, \qquad \text{or,}$$

$$P = \frac{2B'^2}{A'}.$$

*Scholium* 5. The parameter of the transverse axis is equal to

$$\frac{2B^2}{A}.$$

and of the conjugate axis to $\dfrac{2A^2}{B}$

It may be easily shown that the chord drawn through the focus and perpendicular to the transverse axis, is equal to the parameter of that axis.

*Scholium* 6. If through the extremities of any diameter two supplementary chords be drawn, they will enjoy analogous properties to those drawn through the vertices of the transverse axis.

Let $AB$ be any diameter and designate it by $2A'$. Let the axis of $X$ coincide with this diameter, and the axis of $Y$ with the conjugate diameter $DD'$. Designate the angle $DCB$ by $\beta$. Then, if through $B$ whose co-ordinates are $y' = 0$, and $x' = A'$, a right line be drawn, making with $AB$ an angle equal to $\alpha$ its equation will be of the form,

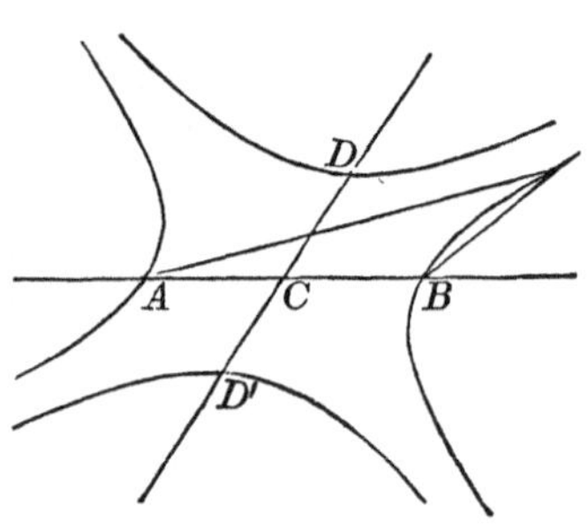


$$y = a(x - A') \qquad a = \frac{\sin \alpha}{\sin(\beta - \alpha)}.$$

If through $A$ whose co-ordinates are $y' = 0$, $x' = \ A'$

a second right line be drawn, making with $AB$ an angle $\alpha'$, we shall have,

$$y = a'(x + A'), \qquad a' = \frac{\sin \alpha'}{\sin(\beta - \alpha')}.$$

Combining these equations with each other and with the equation of the hyperbola, we obtain

$$A'^2 aa' - B'^2 = 0, \qquad \text{or,} \qquad aa' = \frac{B'^2}{A'^2}.$$

for the equation of condition when the lines are supplementary chords.

*Scholium* 7. If it be required to draw a tangent line to the curve at any point of which the co-ordinates are $x''$, $y''$, we must combine the three equations (Prop. IV),

$$A'^2y^2 - B'^2x^2 = -A'^2B'^2,$$

$$A'^2y''^2 - B'^2x''^2 = -A'^2B'^2,$$

$$y - y'' = \frac{y' - y''}{x' - x''}(x - x''),$$

which will give for the equation of the tangent,

$$A'^2yy'' - B'^2xx'' = -A'B'^2.$$

*Scholium* 8. If through the centre of the hyperbola and the point of contact a diameter be drawn, its equation will be

$$y = a'x,$$

or,

$$a' = \frac{y''}{x''}.$$

But

$$a = \frac{B'^2x''}{A'^2y''}.$$

Hence,

$$aa' = \frac{B'^2}{A'^2}.$$

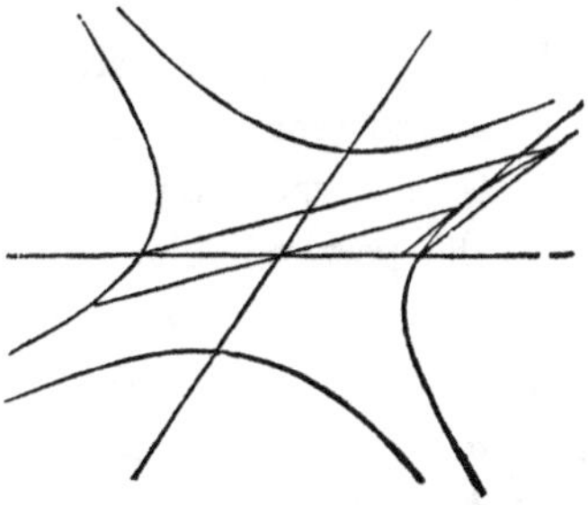

Comparing this with the equation of condition of supplementary chords, we see, that *two supplementary chords may always be drawn respectively parallel to a diameter and a tangent line through its vertex.*

*Scholium* 9. If we resume the equation of the hyperbola referred to its centre and conjugate diameters, which is

$$A^{2\prime}y^2 - B^{\prime 2}x^2 = -A^{\prime 2}B^{\prime 2},$$

and then refer it to its centre and axes, and compare the coefficients with those of the known equation,

$$A^2y^2 - B^2x^2 = -A^2B^2,$$

we shall obtain after adding two of the equations as in (Bk. IV, Prop. XI),

$$A'B' \sin(\alpha' - \alpha) = AB. \qquad (1)$$

$$A'^2 - B'^2 = A^2 - B^2. \qquad (2);$$

or, these equations may be obtained directly from the corresponding equations of the ellipse by substituting for $B$, $B\sqrt{-1}$, and for $B'$, $B'\sqrt{-1}$.

*Scholium* 10. The first of these equations shows, that *the parallelogram formed by drawing tangent lines through the vertices of conjugate diameters, is equivalent to the rectangle formed by drawing tangent lines through the vertices of the axes.*

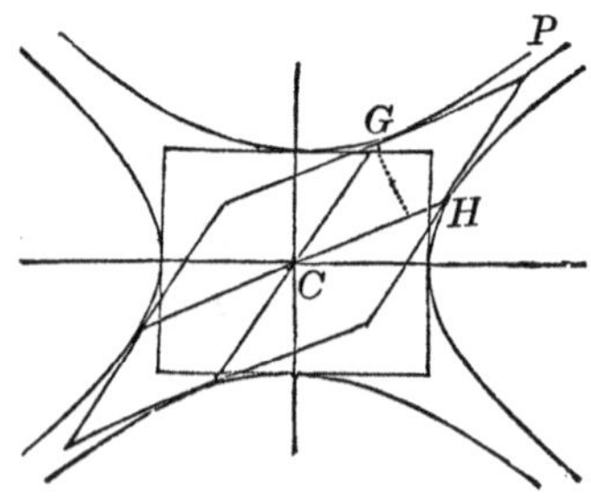


Having formed the parallelogram and rectangle, draw from $G$ a perpendicular to $CH$, this perpendicular will be equal to $B' \sin(\alpha' - \alpha)$. Hence, the area of the parallelogram $CGPH$ is equal to $A'B' \sin(\alpha' - \alpha) = AB$.

therefore the whole parallelogram is equivalent to the whole rectangle.

*Scholium* 11. The second equation,

$$A'^2 - B'^2 = A^2 - B^2,$$

or,

$$4A'^2 - 4B'^2 = 4A^2 - 4B^2,$$

expresses that, *the difference of the squares of two conjugate diameters is equal to the difference of the squares of the axes.*

Hence, there can be no equal conjugate diameters unless $A = B$, in which case the hyperbola is equilateral, and then, *every diameter will be equal to its conjugate.*

## *Of the Hyperbola referred to its Asymptotes.*

If the diagonals of the rectangle described on the axes of the hyperbola be indefinitely produced in both directions, the lines so drawn are called *asymptotes* of the hyperbola.

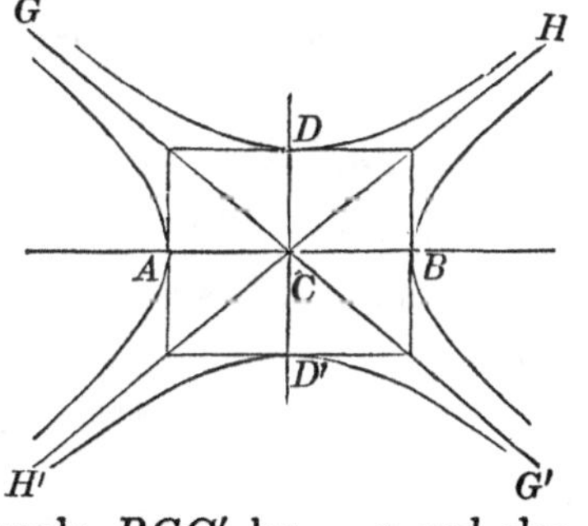


Thus, $H'H$, $G'G$, are asymptotes of the hyperbola whose transverse axis is $AB$, and also of the conjugate hyperbola whose transverse axis is $DD'$.

If we designate the angle estimated from $CB$ around to $CG'$ by $\alpha$, or do what is equivalent to it, designate the angle $BCG'$ by $-\alpha$, and also represent the angle $BCH$ by $\alpha'$, we shall have,

$$\text{tang}\ \alpha = -\frac{B}{A}, \qquad \text{tang}\ \alpha' = \frac{B}{A},$$

or

$$\text{tang}^2\alpha = \frac{B^2}{A^2}, \qquad \text{tang}^2\alpha' = \frac{B^2}{A^2},$$

or

$$A^2 \sin^2\alpha - B^2 \cos^2\alpha = 0, \qquad A^2 \sin^2\alpha' - B^2 \cos^2\alpha' = 0$$

which are the two equations expressing the relations between the angles which the asymptotes form with the transverse axis.

These equations may be placed under the forms,

$$\sin \alpha = \frac{\mp B}{\sqrt{A^2+B^2}}, \qquad \sin \alpha' = \frac{\pm B}{\sqrt{A^2+B^2}}$$

$$\cos \alpha = \frac{\pm A}{\sqrt{A^2+B^2}}, \qquad \cos \alpha' = \frac{\pm A}{\sqrt{A^2+B^2}},$$

by substituting for $\cos^2$, $1-\sin^2$, and for $\sin^2$, $1-\cos^2$.

If $A = B$, the hyperbola is equilateral, and we shall then have,

$$\sin \alpha = -\cos \alpha, \qquad \text{and} \qquad \sin \alpha' = \cos \alpha',$$

which indicates that the asymptotes lie on different sides of the transverse axis, and make angles of 45° with it. Hence,

*In the equilateral hyperbola the asymptotes are at right angles to each other.*

## PROPOSITION IX. PROBLEM.

*To find the equation of the hyperbola referred to its centre and asymptotes.*

The equation of the hyperbola referred to its centre and axes is,

$$A^2y^2 - B^2x^2 = -A^2B^2.$$

The formulas for passing from rectangular to oblique co-ordinates, the origin remaining the same, are

$$x = x' \cos \alpha + y' \cos \alpha', \qquad y = x' \sin \alpha + y' \sin \alpha'$$

Substituting these values and reducing, we obtain

$$\left.\begin{matrix}(A^2\sin^2\alpha' - B^2\cos^2\alpha')y'^2 + (A^2\sin^2\alpha - B^2\cos^2\alpha)x'^2 \\ + 2(A^2\sin\alpha\sin\alpha' - B^2\cos\alpha\cos\alpha')x'y'\end{matrix}\right\} = -A^2B^2$$

The equations of condition reduce the coefficients of $x'^2$, and $y'^2$ to 0, and that of $x'y'$ to

$$-\frac{4A^2B^2}{A^2+B^2}:$$

hence, the equation of the hyperbola referred to its asymptotes becomes

$$x'y'=\frac{A^2+B^2}{4},$$

or by putting $M$ for $\frac{A^2+B^2}{4}$, and omitting the accents

$$xy=M.$$

*Scholium* 1. *The curve of the hyperbola continually approaches the asymptotes, and becomes tangent to them at an infinite distance from the centre.*

For, the equation of the hyperbola referred to its asymptotes gives

$$y=\frac{M}{x}.$$

Now, since $M$ is constant, if $x$ increases continually $y$ will diminish, and if $x$ becomes infinite, $y$ will become 0: hence, the hyperbola continually approaches the asymptote, and as $y$ cannot become negative so long as $x$ is positive, it follows that the curve will touch the asymptote when $y$ is 0. The same might be shown with respect to the axis of $Y$.

*Scholium* 2. It may also be easily shown, that the asymptotes are the limits of all straight lines drawn tangent to the hyperbola.

The equation of the tangent, referred to the axes, is

$$A^2yy''-B^2xx''=-A^2B^2.$$

If we make $y=0$, we find

$$x=\frac{A^2}{x''},$$

which is the distance from the centre to the point in which the tangent intersects the transverse axis.

If now, $x''$ be made infinite, $x$ will be equal to 0, that is, the tangent line will pass through the centre, and since both the tangent and asymptote touch the curve at a point infinitely distant from the centre, they will coincide.

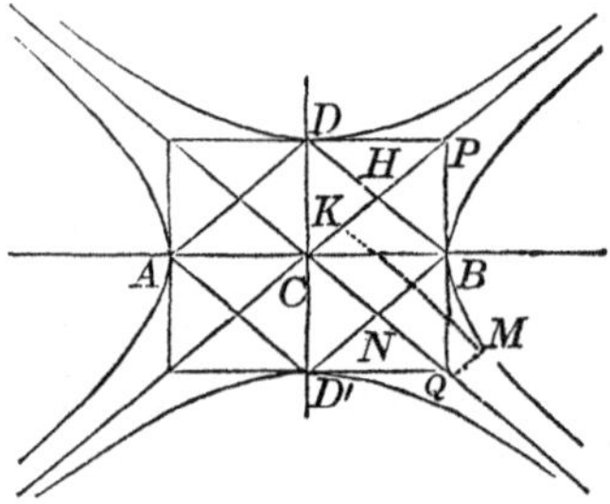


If lines be drawn through the vertices of the axes, they will form the rhombus $AD'BD$. The diagonals $CP$, $CQ$, of the rectangles described on the semi-axes, are equal to each other, and each is equal to $\sqrt{A^2+B^2}$ But these diagonals are also equal to $BD$, $BD'$, and each pair of the equal diagonals mutually bisect each other at $H$ and $N$. Hence, $CH=\frac{1}{2}\sqrt{A^2+B^2}$, and $CN=\frac{1}{2}\sqrt{A^2+B^2}$;

$$\text{therefore,} \quad CH\times CN=\frac{A^2+B^2}{4}=xy.$$

If we designate the angle included between the asymptotes by $\beta$, we shall have

$$CH\times CN\sin\beta=xy\sin\beta;$$

the first member of the equation is equal to the rhombus $CHBN$, and the second, to any parallelogram, as $CQMK$, the sides of which are $x$ and $y$; that is,

*The rhombus described on the abscissa and ordinate of the vertex of the hyperbola, is equivalent to the parallelogram described on the abscissa and ordinate of any point of the curve.*

*Scholium* 3. The rhombus $CHBN$, described on the abscissa and ordinate of the vertex of the hyperbola, is called the *power of the hyperbola.*

*Scholium* 4. The rhombus $AD'BD$ is equal to one-half of the rectangle described on the axes of the hyperbola. But the power is equal to one-fourth of this rhombus ; hence, the power of an hyperbola is equal to one-eighth of the rectangle described on the axes.

PROPOSITION X. PROBLEM.

*To find the equation of a tangent line to an hyperbola referred to its centre and asymptotes.*

Let $P$ be the point at which the tangent is to be drawn.

Designate its co-ordinates by $x''$, $y''$. The equation of a line passing through this point is of the form

$$y - y'' = a(x - x'');$$

and it is required to find $a$ when the line becomes tangent to the curve.

The equation of the hyperbola referred to its asymptotes, is

$$xy = M;$$

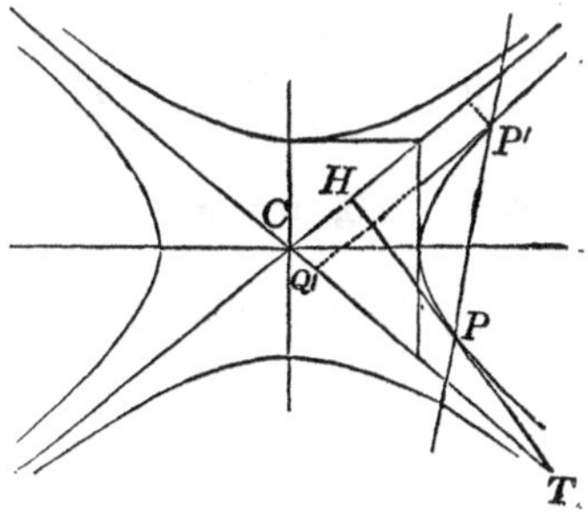

and since $P$ is on the curve,

$$x''y'' = M.$$

Subtracting the last equation from the preceding, we obtain

$$xy - x''y'' = 0,$$

which may be put under the form

$$x(y - y'') + y''(x - x'') = 0.$$

13

Combining this with the equation of the secant line, and we obtain

$$ax(x-x'')+y''(x-x'')=0,$$

or

$$(x-x'')(ax+y'')=0.$$

This equation will be satisfied by making $x=x''$, which will give $y=y''$, and this will designate the first point $P$, in which the secant cuts the curve. The abscissa of the second point $P'$ will be

$$x=-\frac{y''}{a}=CQ'.$$

If we make the second point coincide with the first, we shall again have $x=x''$, and $y=y''$, and this will give

$$a=-\frac{y''}{x''}.$$

Substituting this value, and we have

$$y-y''=-\frac{y''}{x''}(x-x''),$$

which is the equation of the tangent line $HT$.

*Scholium* 1. If in the equation of the tangent

$$y-y''=-\frac{y''}{x''}(x-x''),$$

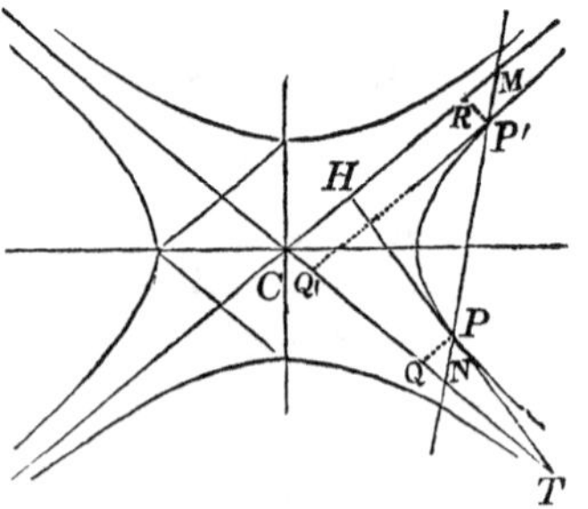


we make $y=0$, we find

$$CT=x=2x'',$$

or

$$x-x''=x''.$$

But $x-x''$ is equal to the sub-tangent $QT$: hence, *the sub-tangent referred to the asymptotes, is equal to the abscissa of the point of tangency.*

*Scholium* 2. Since $CQ$ is equal to $QT$, it follows, from similar triangles, that $TP$ is equal to $PH$: hence,

*If at any point of an hyperbola a line be drawn tangent to the curve, the part of the tangent intercepted by the asymptotes will be bisected at the point of tangency.*

*Scholium* 3. If in the equation of the secant $NM$, which is

$$y - y'' = a(x - x''),$$

we make $y = 0$, we shall have $x = CN$, and

$$x - x'' = -\frac{y''}{a} = QN.$$

But in determining the equation of the tangent, we found

$$CQ' = -\frac{y''}{a}:$$

hence, $$NQ = CQ'.$$

If then, $P'R$ be drawn parallel to the asymptote $CT$, the triangles $NPQ$, $P'RM$ will be similar: and since the bases $NQ$, $P'R$ are equal, $PN$ will be equal to $P'M$. Hence,

*If a line be drawn meeting the asymptotes and cutting the hyperbola in two points, the distance from either of the points to one of the asymptotes will be equal to the distance from the other point to the other asymptote.*

*Scholium* 4. The last property affords an easy method of describing an hyperbola by points, when the asymptotes and one point of the curve are known.

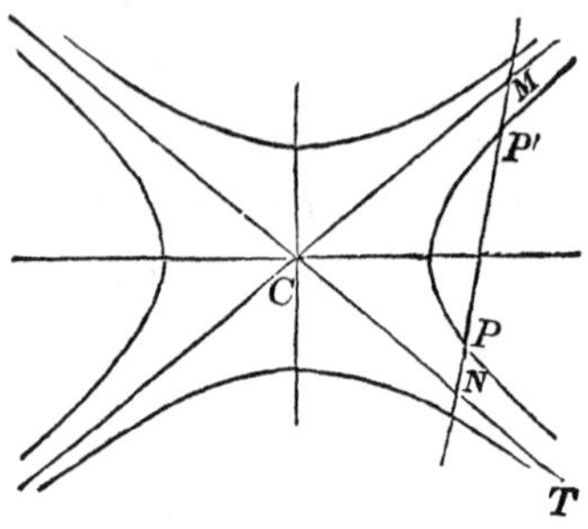


Let $CT$, $CM$, be the asymptotes of a hyperbola, and $P$ a point of the curve.

Through $P$ draw any right line, as $NPM$, cutting the asymptotes in $N$ and $M$.

Take the distance $PN$, which is known, and lay it off from $M$ to $P'$: then will $P'$ be a point of the curve. In a similar manner any number of points may be found.

PROPOSITION XI. THEOREM.

*If a tangent line be drawn to the hyperbola, and limited by the asymptotes, it will be equal to the conjugate of the diameter which passes through the point of contact.*

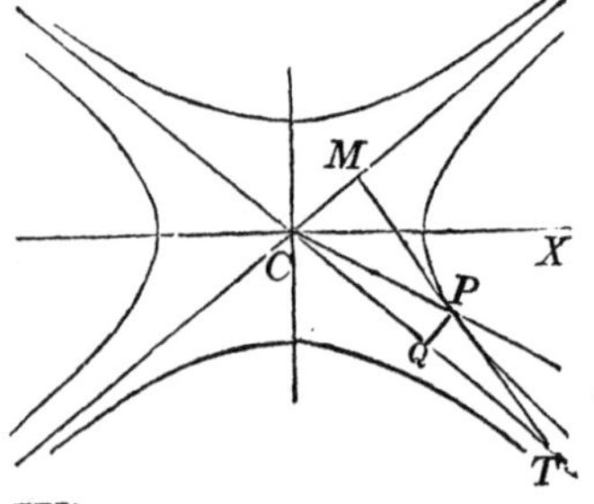


Let $TM$ be the tangent line touching the hyperbola at $P$.

Through $P$ draw the semi-diameter $CP$. Denote the semi-diameter $CP$ by $A'$, and the angle $MCT$, formed by the asymptotes, by $\beta$.

The two triangles, $CPQ$, $QPT$, will then give (Trig. Th. IV),

$$\overline{CP}^2 = x^2 + y^2 + 2xy \cos \beta,$$

$$\overline{TP}^2 = x^2 + y^2 - 2xy \cos \beta.$$

Hence, $$\overline{CP}^2 - \overline{PT}^2 = 4xy \cos \beta.$$

But, since the angle $MCX = XCT$, we have $\beta = 2\alpha'$ ·

hence $$\cos \beta = \cos^2\alpha' - \sin^2\alpha' \quad \text{(Trig. Art. XX)}.$$

But the equations of condition give,

$$\cos^2\alpha' = \frac{A^2}{A^2+B^2}, \quad \text{and} \quad \sin^2\alpha' = \frac{B^2}{A^2+B^2},$$

hence, $$\cos\beta = \frac{A^2-B^2}{A^2+B^2}.$$

The equation of the hyperbola, referred to its asymptotes, also gives

$$4xy = A^2 + B^2:$$

hence, $$\overline{CP}^2 - \overline{PT}^2 = A^2 - B^2,$$

or $$A'^2 - \overline{PT}^2 = A^2 - B^2.$$

Now, it has already been proved, that the difference of the squares of the two conjugate diameters is equal to the difference of the squares of the axes (Prop. VIII, Sch. 11).

But $A$ and $B$ are the semi-axes, and $A'$ is a semi-diameter hence, $PT$ must be equal to the semi-conjugate; and therefore the tangent $TM$ is equal to the diameter which is conjugate to $CP$.

*Scholium*. Let $H'H$, $G'G$ be two conjugate diameters, and through their vertices let tangent lines be drawn, forming a parallelogram.

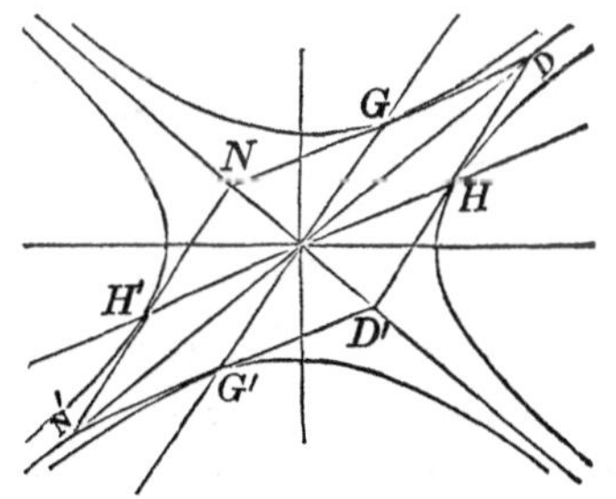


Then, since the tangents $D'D$, $NN'$, are equal and parallel to the diameter $G'G$, and the tangents $ND$, $N'D'$ equal and parallel to $H'H$, the vertices of the parallelogram will fall on the asymptotes.

Hence, *the asymptotes are the diagonals of all the parallelograms which can be formed by drawing tangent lines through the vertices of conjugate diameters.*

## *Of the Polar Equation of the Hyperbola.*

### PROPOSITION XII. PROBLEM.

*To find the general polar equation of the hyperbola.*

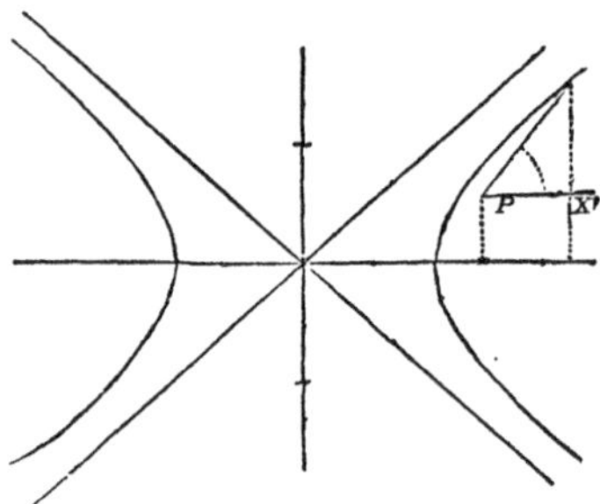


If we designate the co-ordinates of the pole $P$ by $a$ and $b$, and estimate the angles $v$ from the line $PX'$ parallel to the transverse axis, we shall have the following formulas for passing from rectangular to polar co-ordinates: viz.,

$$x = a + r\cos v, \qquad y = b + r\sin v.$$

If we substitute these values of $x$ and $y$, in

$$A^2y^2 - B^2x^2 = -A^2B^2,$$

the equation of the hyperbola referred to its centre and axes, we shall obtain

$$\begin{array}{r|l|l} A^2\sin^2 v & r^2 + 2A^2b\sin v & r + A^2b^2 - B^2a^2 = -A^2B^2, \\ -B^2\cos^2 v & \quad -2B^2a\cos v & \end{array}$$

which is the general polar equation of the hyperbola.

*Scholium* 1. The discussion of the cases in which the pole is placed on the curve, or at the centre, being entirely similar to the corresponding cases of the ellipse, we shall only discuss the equation under the supposition of the pole being placed at either of the foci.

If the pole be placed at the focus on the positive side of abscissas, of which the co-ordinates are

$$a = \sqrt{A^2 + B^2}, \qquad b = 0.$$

the polar equation will become

$$(A^2 \sin^2 v - B^2 \cos^2 v) r^2 - 2 B^2 a \cos v \,.\, r = B^4$$

If we now resolve the equation with reference to $r$, and treat the two roots as in Bk. IV, Prop. XII, Sch. 3, we shall find the two values

$$r = \frac{B^2(a \cos v + A)}{A^2 \sin^2 v - B^2 \cos^2 v}, \qquad r = \frac{B^2(a \cos v - A)}{A^2 \sin^2 v - B^2 \cos^2 v},$$

which, by changing the form of the denominator, and striking out the common factors, reduce to

$$r = \frac{B^2}{A - a \cos v}, \qquad r = -\frac{B^2}{A + a \cos v}.$$

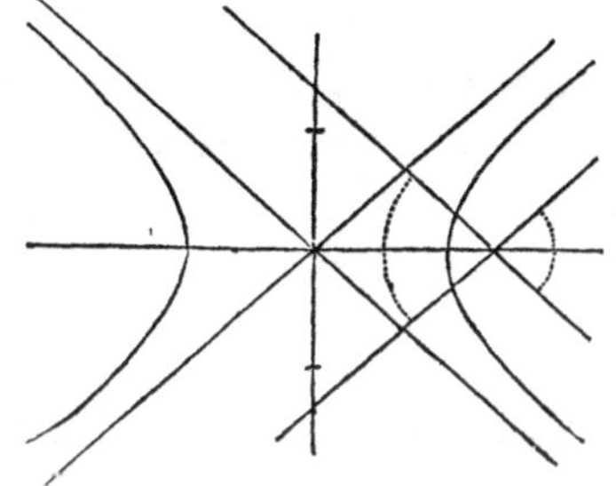

Let us now discuss the first value: that is,

$$r = \frac{B^2}{A - a \cos v}.$$

If we make $v = 0$, we have, $\cos v = 1$, and

$$r = \frac{B^2}{A - a}.$$

But $a = \sqrt{A^2 + B^2}$; therefore the denominator will be negative, and consequently the value of $r$ will be negative: hence there will be no point of the curve for the value of $v = 0$.

The value of $r$ will continue negative for all values of $v$ which give

$$a \cos v > A,$$

and it will be positive for all values of $v$ which give

$$a \cos v < A.$$

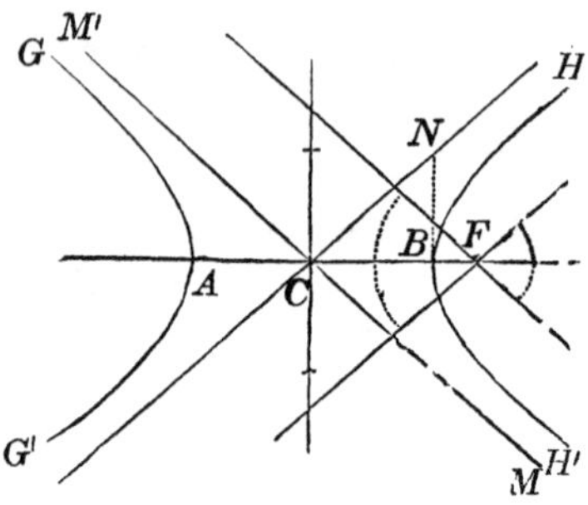

The positive values of $r$ will therefore begin at the point where

$$A = a \cos v,$$

or where $\cos v = \dfrac{A}{\sqrt{A^2 + B^2}}$,

and for that value of $v$ the radius-vector will be infinite.

The cosine of the angle $NCB$ formed by the asymptote with the transverse axis is equal to $CB$ divided by $CN$; that is, to

$$\frac{A}{\sqrt{A^2 + B^2}}.$$

Hence, when the radius-vector through $F$ becomes parallel to the asymptote $CN$, it will be positive and infinite, and will determine that point of the curve at which the asymptote is tangent.

For all values of $v$, of which the cosine is less than

$$\frac{A}{\sqrt{A^2 + B^2}},$$

or of which the cosine is negative, $r$ will be positive, and will therefore give points of the curve.

When the value of $v$ is such that the radius-vector through $F$ becomes parallel to the asymptote $CM$, we shall again have

$$\cos v = \frac{A}{\sqrt{A^2 + B^2}}:$$

hence, $r$ will again become infinite; and from this value of $v$ to $v = 360°$, the value of $r$ will be negative The first value of $r$ will therefore give all the points of the branch $H'BH$ of the hyperbola.

Let us now discuss the second value: viz.

$$r = \frac{-B^2}{A + a \cos v}.$$

This value of $r$ can only be positive when the denominator is negative, which requires that we should have,

$$A < -a \cos v.$$

If then we make

$$-a \cos v = A,$$

or,
$$\cos v = \frac{-A}{\sqrt{A^2 + B^2}},$$

we shall find the limit of the values of $v$ which can render $r$ positive. For this value of $v$ the radius-vector through $F$ is parallel to the asymptote $CM'$, and infinite; and therefore determines the point at which the asymptote is tangent to the branch $G'AG$. The value of $r$ will then continue positive until the radius-vector becomes parallel to the asymptote $CN'$, when it again becomes infinite and then negative. The second value of $r$, therefore, determines the branch $G'AG$. The first value therefore answers to the case in which the pole is placed at the focus within the curve, and the second, to that in which it is placed at the focus without the curve. *The two together give positive values of* r *for an angular space of* 360°.

*Scholium* 2. If, as in the ellipse, we make

$$e = \frac{a}{A} = \frac{\sqrt{A^2 + B^2}}{A},$$

we shall have, for the polar equation when the pole is within the curve

$$r = -\frac{A(1 - e^2)}{1 - e \cos v},$$

and for the polar equation when the pole is without the curve

$$r = +\frac{A(1 - e^2)}{1 + e \cos v}.$$

in both of which equations the numerator is equal to half the parameter of the transverse axis.

*General Scholium*

1. We have seen (Bk. IV, Prop. I, Sch. 8), that if the origin of co-ordinates be placed at the vertex of the transverse axis, the equation of the ellipse will be

$$y^2 = \frac{B^2}{A^2}(2Ax - x^2);$$

the equation of the parabola for a similar position of the origin (Bk. V, Prop. I), is

$$y^2 = 2px,$$

and for that of the hyperbola (Prop. I, Sch. 7),

$$y^2 = \frac{B^2}{A^2}(2Ax + x^2).$$

These equations may all be put under the form

$$y^2 = mx + nx^2,$$

in which $m$ is the parameter of the curve, and $n$ the square of the ratio of the semi-axes. In the ellipse $n$ is negative; in the hyperbola it is positive, and in the parabola it is 0.

2. The curves whose properties have been discussed in the three last books are precisely those which are obtained by intersecting the surface of a cone by planes, as is shown in (Bk. IX, Art. 21). For this reason they are called *Conic Sections.*

3. There is a general property of these curves too important not to be particularly noted. It is this: *If the pole be placed at the focus, the radius-vector will always be expressed rationally in terms of the abscissa of the point in which it intersects the curve.*

In page 97, we have, for the ellipse

$$r' = A + \frac{cx}{A} \quad \text{and} \quad r = A - \frac{cx}{A}.$$

In page 141, we have, for the parabola

$$FP = r = \frac{p}{2} + x.$$

In page 166, we have, for the hyperbola

$$r' = A + \frac{cx}{A} \quad \text{and} \quad r = -A + \frac{cx}{A},$$

in all of which, the value of $r$ is expressed rationally in $x$. It can be rigorously proved that the focus is the only point in the plane of the curve which enjoys this property.

# BOOK VII.

## *Discussion of the General Equation of the second degree between two Variables.*

1. It has been shown that every equation of the first degree, between two variables, is the equation of a straight line (Bk. II, Prop. II).

We have also seen, that the equation of the circle, the equation of the ellipse, the equation of the parabola and the equation of the hyperbola, are all of the second degree; and analogy might lead us to infer, that *every equation of the second degree between two variables*, must represent one or the other of these curves. This is what we now propose to prove rigorously.

The general equation of the second degree between two variables, is

$$Ay^2 + Bxy + Cx^2 + Dy + Ex + F = 0,$$

which contains the first and second powers of each variable, their product, and an absolute term $F$.

The coefficients, $A$, $B$, $C$, $D$, $E$, and $F$, are entirely independent of the variables $y$ and $x$, and values may be assigned to them at pleasure; but when once assigned, those values remain constant throughout the same discussion.

These coefficients are called *constants;* but this by no means implies that they always retain the same value, for indeed, *the discussion of the equation consists in tracing out all the changes to which it is subjected, by the different suppositions which can be made on the absolute and relative values of these coefficients.*

2. Let us suppose, in the first place, that the co-ordinate axes are rectangular. This supposition will not render the discussion and the results less general. For, if the co-ordinate axes were oblique, we might readily pass to a system of rectangular co-ordinates, without affecting the degree of the equation, since the equations for transformation are always linear, or of the first degree.

3. Let us begin the discussion by supposing, that

$$A = 0, \qquad \text{and} \qquad C = 0.$$

The general equation will then become

$$Bxy + Dy + Ex + F = 0.$$

If now we refer the curve to a system of parallel axes, of which the co-ordinates of the new origin, with reference to the primitive axes, are $a$ and $b$, the formulas are (Bk. II Prop. X),

$$x = a + x', \qquad y = b + y'.$$

These values of $x$ and $y$ being substituted in the previous equation, it will become

$$Bx'y' + (Ba + D)y' + (Bb + E)x' + Bab + Db + Ea + F = 0.$$

The co-ordinates of the new origin may be regarded as undetermined, and such values may be attributed to them as shall cause the equation to take a particular form.

Let us then make

$$Ba + D = 0 \qquad \text{and} \qquad Bb + E = 0,$$

which gives $$a = -\frac{D}{B}, \qquad b = -\frac{E}{B};$$

and these values will reduce the equation in $x'y'$, to the form

$$Bx'y' - \frac{DE}{B} + F = 0, \quad \text{or} \quad x'y' = \frac{DE - BF}{B^2};$$

and since the axes are at right angles, this is the equation of an equilateral hyperbola referred to its centre and asymptotes and of which $\frac{DE - BF}{B^2}$ is the power.

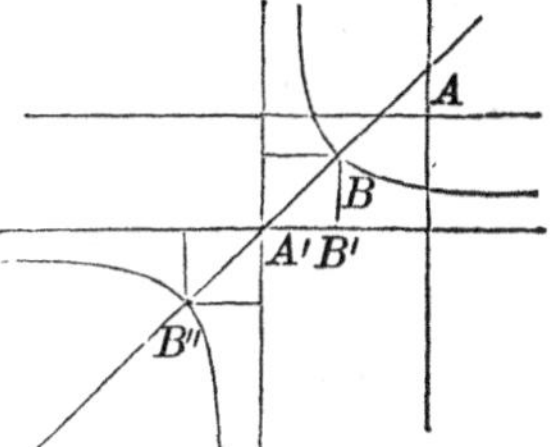


To construct this equation, let us suppose $A$ to be the origin of the primitive axes. From $A$ lay off, on the negative part of the axis of $X$, a distance equal to $-\frac{D}{B}$. Lay off also on the negative part of the axis of $Y$, a distance equal to $-\frac{E}{B}$, and through the points draw parallels to the axes, their point of intersection $A'$, will be the new origin.

The line drawn through the origin $A'$, and bisecting the angle of the co-ordinate axes, will be the transverse axis of the curve.

If we make $A'B'$ equal to $\frac{\sqrt{DE - BF}}{B}$, and draw $B'B$ parallel to the other asymptote, $B$ will be the vertex of the transverse axis. The other vertex, $B''$, is determined in the same manner.

4. Let us suppose that we make, at the same time,

$$A = 0, \qquad B = 0, \qquad C = 0:$$

the general equation will then reduce to

$$Dy + Ex + F = 0,$$

which is the equation of a straight line.

5. These particular cases offer no difficulty, and may therefore be excluded from the general discussion. Let us, therefore, suppose that the second power of at least one of the variables, $y$ for example, enters into the equation.

6. Resolving the general equation with reference to $y$ we obtain,

$$y = -\frac{1}{2A}(Bx+D) \pm \frac{1}{2A}\sqrt{(B^2-4AC)\,x^2+2(BD-2AE)x+D^2-4AF}$$

This value of $y$ is composed of two distinct parts: the one

$$-\frac{1}{2A}(Bx+D),$$

and the other the radical part of the second member.

Since, $$y = -\frac{1}{2A}(Bx+D) = -\frac{B}{2A}x - \frac{D}{2A},$$

is the equation of a straight line, the second member, which represents the first part of the general value of $y$, may easily be constructed.

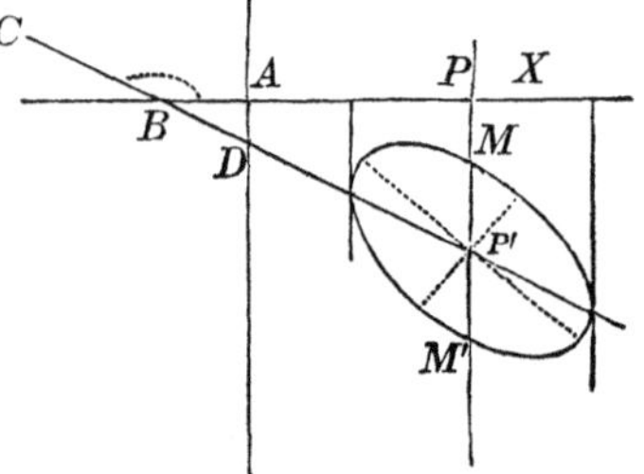


Let $A$ be the origin of a system of rectangular axes.

Lay off from $A$ a distance $AD$, in the negative direction of the ordinates, equal to

$$-\frac{D}{2A};$$

and through $D$ draw $DBC$, making with the axis of $X$ an angle $XBC$, whose tangent shall be equal to $-\frac{B}{2A}$. The angle will be obtuse, since its tangent is negative.

The ordinate of any point of this line will be that part of the value of $y$ which is without the radical sign. The line $CD$ has been drawn under the supposition that $-\frac{B}{2A}$, is essentially negative, which requires that $B$ and $A$ should have the same sign. If the coefficient of $x$ were positive the angle $XBC$ should be made acute

If now we lay off from the origin $A$ any abscissa, as $AP$, and designate it by $x$, we shall have $PP'$ for that part of the corresponding value of $y$ which is without the radical. For, from the equation

$$y = -\frac{1\,Bx}{2A} - \frac{D}{2A},$$

if $x = AP$, we shall have $y = PP'$.

If we now lay off from $P'$, the plus value of the radical from $P'$ to $M$, above $CP'$, and the minus value from $P'$ to $M'$, below $CP'$, $PM$ will represent one value of $y$, and $PM'$ the other. The corresponding values of the ordinates may be determined in a similar manner for any abscissa whatever.

Since all the points of the curve are determined by laying off from the different points of $CP'$ two equal lines in a contrary direction, it follows that the curve will be symmetrical with respect to the line $CP'$: that is, $CP'$ will bisect a system of parallel chords which terminate in the curve: hence, $CP'$ is a diameter. And generally, *if an equation of the second degree between two variables be resolved with reference to one of the variables, the first member, together with that part of the second member which is independent of the radical, will be the equation of a diameter of the curve.*

7. Since the curve is symmetrical with respect to the diameter $CP'$, its equation may be simplified by referring it to this line as an axis of abscissas. The origin of co-ordinates will then be transferred to $D$, and the new ordinates $y'$ will be estimated from $CP'$, and parallel to the primitive axis $Y$ Designate by $\alpha$ the angle $XBC$: we shall have

$$\text{tang } \alpha = -\frac{B}{2A},$$

and the equations for transformation will become,

$$x = -x' \cos \alpha, \qquad y = -\frac{1}{2A}(Bx + D) + y'.$$

Substituting these values for $x$ and $y$ in the resolved equation, and we obtain

$$y' = \pm\frac{1}{2A}\sqrt{(B^2-4AC)\cos^2\alpha\, x'^2 - 2(BD-2AE)\cos\alpha\, x' + D^2 - 4AF};$$

squaring both members, and clearing the denominator, we obtain

$$4A^2y'^2 = (B^2-4AC)\cos^2\alpha\,.\,x'^2 - 2(BD-2AE)\cos\alpha\,.\,x' + D^2 - 4AF.$$

The polynomial which forms the second member of this equation, may be placed under the form,

$$(B^2-4AC)\cos^2\alpha\left\{x'^2 - 2\frac{(BD-2AE).x'}{(B^2-4AC)\cos\alpha}\right\} + D^2 - 4AF,$$

in which we see that the variable quantity within the parenthesis will become a perfect square if we add,

$$\frac{(BD-2AE)^2}{(B^2-4AC)^2\cos^2\alpha}.$$

We must, however, to preserve the equality, subtract

$$(B^2-4AC)\cos^2\alpha\left\{\frac{(BD-2AE)^2}{(B^2-4AC)^2\cos^2\alpha}\right\}$$

which reduces to

$$\frac{(BD-2AE)^2}{B^2-4AC}.$$

Making these transformations, we obtain

$$4A^2y'^2 = (B^2-4AC)\cos^2\alpha\left\{x' - \frac{(BD-2AE)}{(B^2-4AC)\cos\alpha}\right\}^2 - \frac{(BD-2AE)^2}{B^2-4AC} + D^2 - 4AF.$$

Let us now transfer the origin of co-ordinates from $D$ to $P'$, making

$$DP'=\frac{BD-2AE}{(B^2-4AC)\cos\alpha}.$$

If we continue the new axis of ordinates parallel to the primitive, and designate the co-ordinates referred to the new origin by $x''$, $y''$, the equations for transformation will be,

$$x'=\frac{BD-2AE}{(B^2-4AC)\cos\alpha}+x'', \qquad \text{and} \qquad y'=y''$$

Substituting these values, and the equation of the curve referred to the new system of co-ordinates, will become,

$$4A^2y''^2=(B^2-4AC)\cos^2\alpha x''^2-\frac{(BD-2AE)^2}{B^2-4AC}+D^2-4AF,$$

$$\text{or,}\ 4A^2y''^2-(B^2-4AC)\cos^2\alpha x''^2=-\frac{(BD-2AE)^2}{B^2-4AC}+D^2-4AF.$$

This equation contains but the second powers of the variables with coefficients, and an absolute term which forms the second member.

The coefficient of $y''^2$ is positive, and the sign of the coefficient of $x''^2$ depends on the sign of $B^2-4AC$, since the $\cos^2\alpha$ is positive.

This equation will take the form of the equation of an ellipse referred to its centre and conjugate diameters, when $B^2-4AC$ is negative: for then the essential sign of the coefficients of the second powers of the variables will be both positive.

8. If we make $B=0$, and $A=C$, the essential sign of the coefficient of $x''^2$ will still be positive. Under this supposition the coefficients of $y''^2$, $x''^2$, will become equal to each other, and the equation will take the form

$$x''^2+y''^2=R^2.$$

Hence, the equation of the circle belongs to the same general class with that of the ellipse, and may be derived from it by making particular suppositions on the *values* of the constants. This is as it should be, since the ellipse becomes the circle by making the axes equal to each other.

9. If $B^2 - 4AC$ be positive, the equation will take the form of the equation of a hyperbola referred to its centre and axes.

10. In transferring the origin from $D$ to $P'$, we made

$$DP' = \frac{(BD - 2AE)}{(B^2 - 4AC)\cos\alpha}.$$

If $DP'$ becomes infinite, the transformation is obviously impossible. When this occurs,

$$(B^2 - 4AC)\cos\alpha = 0,$$

which requires that

$$B^2 - 4AC = 0, \qquad \text{or} \qquad \cos\alpha = 0.$$

But we have made

$$\text{tang}\,\alpha = -\frac{B}{2A},$$

and as $y^2$ enters into the equation, $A$ cannot be 0; hence, tang $\alpha$ cannot be infinite; therefore, cos $\alpha$ cannot be 0: hence, the supposition requires that

$$B^2 - 4AC = 0.$$

Introducing this condition into the second transformed equation, and it becomes

$$4A^2y'^2 = -2(BD - 2AE)\cos\alpha . x' + D^2 - 4AF,$$

or $$y'^2 = -\frac{2(BD - 2AE)}{4A^2}\cos\alpha . x' + \frac{D^2 - 4AF}{4A^2},$$

or $$y'^2 = Px' + Q,$$

by representing the coefficient of $x'$ by $P$, and the absolute term by $Q$.

If now we transfer the origin of co-ordinates in the direction of the negative abscissas, and to a point at a distance from the origin equal to $\frac{Q}{P}$, and continue the new axis of $Y$ parallel to the primitive, the equations of transformation will be

$$x' = -\frac{Q}{P} + x'', \quad \text{and} \quad y' = y''.$$

Substituting these values and the equation reduces to

$$y''^2 = Px'',$$

which is the equation of the parabola referred to a system of co-ordinates having their origin at the vertex of a diameter.

We therefore see that, by attributing proper values and signs to the constants, the general equation of the second degree may be made to represent, in succession, all the lines of the second order which have been discussed; and that it cannot represent any others.

We also see, that the lines of the second order are divided into three classes, of which the following are the analytical characteristics:

| | |
|---|---|
| For the ellipse, | $B^2 - 4AC < 0$; |
| for the parabola, | $B^2 - 4AC = 0$; |
| for the hyperbola, | $B^2 - 4AC > 0$. |

We shall discuss these classes in succession.

## *Of the Ellipse.*

$$B^2 - 4AC < 0.$$

**11.** Let us resume the value of $y$ in the general equation,

$$y = -\frac{Bx+D}{2A} \pm \frac{1}{2A}\sqrt{(B^2-4AC)x^2+2(BD-2AE)x+D^2-4AF}.$$

In the first place construct $CM'$, the diameter of the curve, of which the equation is

$$y = -\frac{Bx + D}{2A}.$$

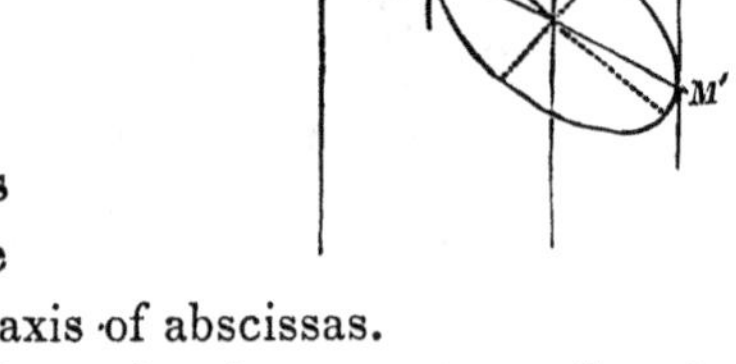


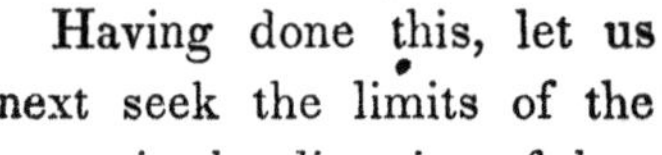

Having done this, let us next seek the limits of the curve in the direction of the axis of abscissas.

All the values of $x$ which render the quantity under the radical sign positive will give real values for $y$, and will therefore correspond to points of the curve.

The values of $x$ which reduce the quantity under the radical to 0, will correspond to those points of the curve which fall on the diameter $CM'$; for such points there will be no values to be laid off above or below $CM'$.

All the values of $x$ which make the quantity under the radical negative, will render $y$ imaginary, and will therefore not correspond to points of the curve.

12. In order to determine the values of $x$ which will satisfy these conditions, let us decompose the polynomial under the radical sign into factors: we may place it under the following form,

$$y = -\frac{Bx+D}{2A} \pm \frac{1}{2A}\sqrt{(B^2-4AC)\left\{x^2+2\frac{(BD-2AE)}{B^2-4AC}x+\frac{D^2-4AF}{B^2-4AC}\right\}}$$

Let us now place

$$x^2 + 2\frac{(BD-2AE)}{B^2-4AC}x + \frac{D^2-4AF}{B^2-4AC} = 0.$$

If we designate the roots of this equation by $x'$, $x''$, the value of $y$ may be placed under the following form (Alg. Art. 142),

$$y = -\frac{Bx + D}{2A} \pm \frac{1}{2A}\sqrt{(B^2 - 4AC)(x - x')(x - x'')}$$

Hence, we see that the values of $y$ will be imaginary or real, according as the product of the factors $x-x'$, $x-x''$, is positive or negative; and consequently, the limits of the curve depend on the values of $x'$, $x''$.

In regard to these values, there are three cases:

1st. When the roots are real and unequal, of which the condition is

$$(BD-2AE)^2-(B^2-4AC)(D^2-4AF)>0.$$

2d. When the roots are real and equal, of which the condition is

$$(BD-2AE)^2-(B^2-4AC)(D^2-4AF)=0.$$

3d. When the roots are both imaginary, of which the con dition is

$$(BD-2AE)^2-(B^2-4AC)(D^2-4AF)<0.$$

1st. *When the roots are real and unequal.*

13. When the roots are real and unequal, all values of $x$, greater than $x'$ and less than $x''$, will give contrary signs to the factors $x-x'$, $x-x''$; their product, $(x-x')(x-x'')$, will then be negative, and as $B^2-4AC$ is also negative the quantity

$$(B^2-4AC)(x-x')(x-x'')$$

will be positive, and consequently the values of $y$ will be real.

If we make $x=x'$, or $x=x''$, the radical will vanish, and the two corresponding values of $y$ will be the ordinates of the vertices of the diameter $N'M'$.

If we designate $AN$, the abscissa of the nearest vertex by $x'$, and $AM$, the abscissa of the remote vertex, by $x''$, we shall have

$$N'N=y'=-\frac{Bx'+D}{2A}, \quad \text{and} \quad M'M=y''=-\frac{Bx''+D}{2A}.$$

Finally, all values of $x$ which are less than $x'$ and all values of $x$ which are greater than $x''$, will render the factors $x-x'$, $x-x''$ of the same sign; hence, their product will be positive; and since $B^2-4AC$ is negative, the quantity under the radical

$$(B^2-4AC)(x-x')(x-x'')$$

will be negative, and this supposition will, therefore, render both values of $y$ imaginary.

We see, therefore, from this discussion, that the curve can have no abscissa less than $x'$, nor greater than $x''$. Hence, the two ordinates $NN'$, $MM'$, drawn through the vertices $N'$, $M'$, will limit the curve in the direction of the axis of $X$; and they will also be tangent to the curve at these points, since they may be regarded as secants of which the two points of intersection have united.

14. When the second power of $x$ enters into the proposed equation, we may resolve it with reference to $x$, and determine the limits of the curve in the direction of the axis of $Y$. The two lines which limit the curve in that direction, will be tangent to it, and parallel to the axis of $X$.

15. Having found the abscissas $x'$, $x''$, of the vertices of the diameter $N'M'$, we can readily find the value of this diameter, and also the value of its conjugate.

The ordinates of the vertices are

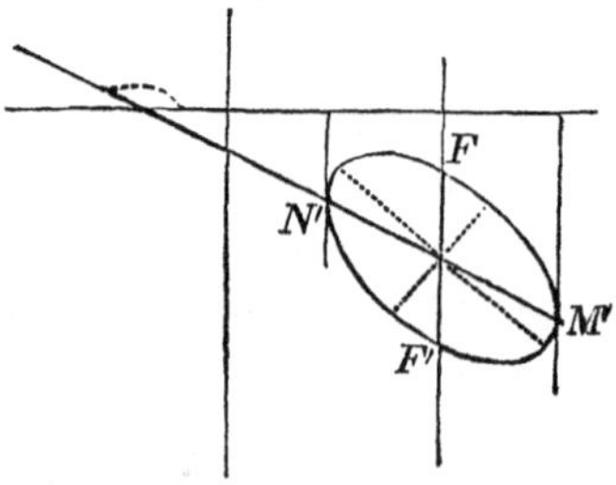

$$y'=-\frac{Bx'+D}{2A},$$

$$y''=-\frac{Bx''+D}{2A};$$

hence, $y''-y'=-\dfrac{B(x''-x')}{2A}$.

But the length of the diameter

is the distance between the two points whose co-ordinates are $x'$, $y'$, $x''$, $y''$: this distance is expressed by

$$\sqrt{(x''-x')^2+(y''-y')^2}.$$

Substituting in this expression the value of $y''-y'$, and we find

$$N'M'=\frac{x''-x'}{2A}\sqrt{B^2+4A^2}.$$

If we designate by $X$ and $Y$ the co-ordinates of the centre of the ellipse, which is the middle point of this diameter, we shall have

$$X=\frac{x'+x''}{2}, \qquad Y=\frac{y'+y''}{2}.$$

If through the centre a diameter be drawn parallel to the axis of $Y$, it will be conjugate to the diameter $N'$ $M'$, since it is parallel to the two tangents drawn through its vertices.

The ordinates of the vertices $F$ and $F'$ of this new diameter, will be the two values of $y$ corresponding to the abscissa of the centre,

$$X=\frac{x'+x''}{2}.$$

If, therefore, we substitute this value for $x$, in the equation

$$y=-\frac{Bx+D}{2A}\pm\frac{1}{2A}\sqrt{(B^2-4AC)(x-x')(x-x'')},$$

we shall obtain

$$y=-\frac{B(x''+x')+2D}{4A}\pm\frac{x''-x'}{4A}\sqrt{4AC-B^2};$$

and the difference of these values will give,

$$FF'=\frac{x''-x'}{2A}\sqrt{4AC-B^2}.$$

16. To find the angle which the conjugate diameters make

with each other, let $\alpha$ designate the angle which the first diameter makes with the axis of $X$. We shall then have

$$\text{tang } \alpha = -\frac{B}{2A}, \qquad \cos \alpha = \frac{2A}{\sqrt{4A^2+B^2}}.$$

Since the second diameter is perpendicular to the axis of $X$, if we designate by $\alpha'$ the angle which it forms with the first, we shall have

$$\alpha' = 90 - \alpha,$$

and

$$\sin \alpha' = \cos \alpha = \frac{2A}{\sqrt{4A^2+B^2}}.$$

Having found the conjugate diameters, and the angle which they make with each other, the ellipse may be described (Bk. IV, Prop. IX, Sch. 4).

17. There are yet other methods of determining points of the curve. If, for example, we wish the points in which the curve intersects the axis of $X$, we make $y = 0$ in the general equation: it will then reduce to

$$Cx^2 + Ex + F = 0.$$

The roots of this equation will be the abscissas of the points common to the curve and the axis of $X$. When the roots are real and unequal, there will be two points of intersection; when they are real and equal the axis of $X$ will be tangent to the curve; and when they are imaginary, the axis will have no point in common with the curve.

18. If in the general equation we make $x = 0$, we shall have

$$Ay^2 + Dy + F = 0;$$

and the roots of this equation will indicate similar relations between the curve and the axis of $Y$.

19. The position of the curve, with respect to the co-ordi nate axes, will then be entirely determined by the values and signs which may be attributed to the constants $A$, $B$, $C$, &c.

The following table exhibits the principal analytical conditions which determine the position of the ellipse with respect to the co-ordinate axes:

| *Positions.* | | *Analytical conditions.* | |
|---|---|---|---|
| For the ellipse | | $B^2-4AC<0$. | (1) |
| Roots $x'$, $x''$, will be real, | when | $(BD-2AE)^2-(D^2-4AF)(B^2-4AC)>0$. | (2) |
| Two points of intersection with the axis $X$, | when | $E^2-4CF>0$. | (3) |
| A point of contact with the axis $X$, | when | $E^2-4CF=0$. | (4) |
| No point of intersection with $X$, | when | $E^2-4CF<0$. | (5) |
| Two points of intersection with the axis of $Y$, | when | $D^2-4AF>0$. | (6) |
| A point of contact with the axis of $Y$, | when | $D^2-4AF=0$. | (7) |
| No point of intersection with the axis of $Y$, | when | $D^2-4AF<0$. | (8) |

20. It is neither easy nor useful to recollect these analytical conditions, but we should not fail to understand clearly the general methods by which they are deduced. The following suggestions may serve as useful guides in the discussion of equations.

1st. When the equation contains the second power of both the variables, resolve it with respect to either of them; but if it contains the second power of but one of the variables, resolve it with respect to that variable.

2d. Construct the diameter of the curve

3d. Place the quantity under the radical sign equal to 0 and the roots of the equation will determine the vertices of the diameter.

4th. Find the points in which the curve intersects the coordinate axes.

The points thus found will, in general, be sufficient to describe the curve.

The following will serve as examples :

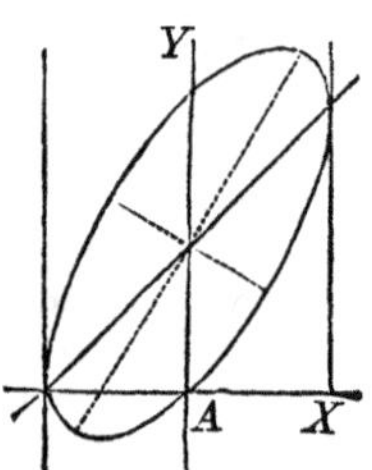


1. $y^2 - 2xy + 2x^2 - 2y + 2x = 0.$

The coefficients of the variables in this equation are such as to satisfy conditions (1), (2), (3) and (6).

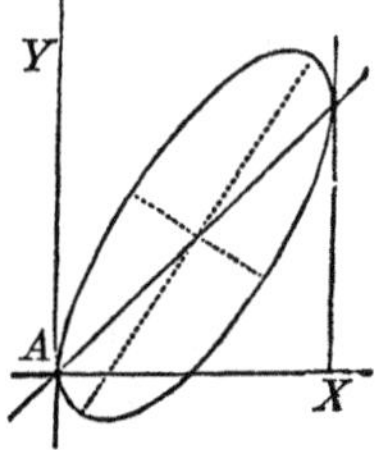


2. $y^2 - 2xy + 2x^2 - 2x = 0.$

This equation will satisfy conditions (1), (2), (3) and (7).

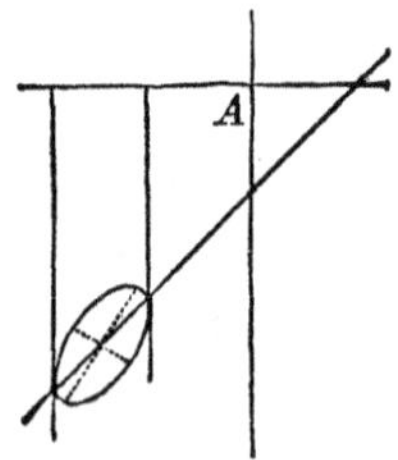


3. $y^2 - 2xy + 2x^2 + 2y + x + 3 = 0.$

This equation will satisfy conditions (1), (2), (5) and (8).

Let us now consider the second case, viz :

2d. *When the roots* $x'$ *and* $x''$ *are equal.*

21. If the roots $x'$, $x''$, are equal to each other, the pro-

duct $(x-x')(x-x'')$ will become the square $(x-x')^2$, and we shall have for the general value of $y$

$$y=-\frac{Bx+D}{2A}\pm\frac{(x-x')}{2A}\sqrt{B^2-4AC}.$$

Now, since $B^2-4AC$ is negative all values of $y$ will be imaginary except the one which corresponds to the value of $x=x'$. This value will cause the radical part to disappear and give

$$y=-\frac{Bx'+D}{2A}.$$

The curve is then reduced to a single point, situated on the diameter, and of which the co-ordinates are

$$x=x' \text{ and } y=-\left\{\frac{Bx'+D}{2A}\right\}$$

By recurring to the equation from which the values of $x'$ and $x''$ were derived, and recollecting that they can only become equal when the quantity under the radical reduces to 0, we shall find

$$(BD-2AE)^2-(B^2-4AC)(D^2-4AF)=0,$$

for the condition which makes $x'=x''$.

This, joined to the condition

$$B^2-4AC<0,$$

reduces the curve to a point.

22. It is easy to show, by reasoning directly upon the general equation, that the condition of $x'=x''$ reduces the curve to a point.

For, we have

$$y=-\frac{Bx+D}{2A}\pm\frac{(x-x')}{2A}\sqrt{B^2-4AC},$$

or $$(2Ay+Bx+D)^2-(x-x')^2(B^2-4AC)=0.$$

But, since $B^2 - 4AC$ is negative, the two terms of the first member of this equation are both positive · hence, the equation can only be satisfied by making

$$2Ay + Bx + D = 0, \qquad \text{and} \qquad x - x' = 0,$$

which gives

$$x = x' \qquad \text{and} \qquad y = -\frac{Bx' + D}{2A},$$

the same conditions as before found.

Now, since $x = x'$ is the only value of $x$ which can satisfy the equation when $x' = x''$, it follows that the substitution of any other value of $x$ will introduce into the equation incompatible conditions, and therefore ought to render the corresponding value of $y$ imaginary.

The following will serve as examples,

$$x^2 + y^2 = 0, \qquad y^2 + x^2 - 2x + 1 = 0.$$

Let us now consider the third case, viz:

3d. *When the roots* x′ *and* x″ *are both imaginary.*

23. Before considering this case we will state a principle of algebra on which its discussion depends, viz:

*When the roots of an equation of the second degree are imaginary, the product of the two binomial factors into which it can be resolved* (Alg. Art. 142), *will be positive*

The roots will be of the form

$$x = \pm a + \sqrt{-b^2}, \qquad x = \pm a - \sqrt{-b^2}$$

and the factors of the form

$$x \mp a - \sqrt{-b^2} \qquad \text{and} \qquad x \mp a + \sqrt{-b^2},$$

and their product

$$(x \mp a - \sqrt{-b^2})(x \mp a + \sqrt{-b^2}) = x^2 \pm 2ax + a^2 + b^2$$

and since $2ax$ cannot exceed $x^2 + a^2$ (Alg. Art. 146), the product will always be positive.

Hence, in the case in which $x'$ and $x''$ are imaginary the product $(x - x')$ $(x - x'')$ will be constantly positive, and since $B^2 - 4AC$ is negative, the quantity under the radical sign will be negative: hence, $y$ will be imaginary for every value of $x$, and consequently there is no curve.

24. In examining the equation from which the value of $x'$ and $x''$ were obtained, we find

$$(BD - 2AE)^2 - (B^2 - 4AC)(D^2 - 4AF) < 0,$$

for the condition which renders $x'$, $x''$, imaginary.

This united with the condition

$$B^2 - 4AC < 0,$$

renders the curve entirely imaginary: that is, *there is no value which can be assigned to either of the variables that will render the corresponding value of the other real.*

The following are equations which give imaginary curves

$$y^2 + xy + x^2 + \frac{1}{2}x + y + 1 = 0, \quad y^2 + x^2 + 2x + 2 = 0,$$

which may be placed under the following forms:

$$(2y + x + 1)^2 + 3x^2 + 3 = 0, \quad y^2 + (x + 1)^2 + 1 = 0.$$

25. In the discussion, the equation has been resolved with reference to $y$. Were we to resolve it with reference to $x$, and place the radical part of the expression equal to 0, we should find for the real and unequal values of $y$,

$$(BE - 2CD)^2 - (B^2 - 4AC)(E^2 - 4CF) > 0;$$

for the equal value of $y$,

$$(BE - 2CD)^2 - (B^2 - 4AC)(E^2 - 4CF) = 0;$$

and for the imaginary curve,

$$(BE - 2CD)^2 - (B^2 - 4CF)(E^2 - 4CF) < 0.$$

But we need not resolve the equation in order to show that if either of these conditions exists for one of the variables, it will necessarily exist for the other.

For, the expression

$$(BD - 2AE)^2 - (B^2 - 4AC)(D^2 - 4AF),$$

includes all the conditions.

If we develop this expression, we shall find it equal to

$$AE^2 + CD^2 + FB^2 - BDE - 4ACF,$$

in which we see, that neither its form nor value will be altered by changing $A$ into $C$, and $D$ into $E$, which is equivalent to changing $y$ into $x$, and $x$ into $y$.

26. There remains yet to be discussed the particular case in which $B = 0$ and $A = C$, or in which the ellipse becomes a circle (Art. 8).

Under this supposition, the general equation reduces to the form

$$A\,y^2 + A\,x^2 + Dy + Ex + F = 0,$$

or by dividing by $A$,

$$y^2 + x^2 + \frac{D}{A}y + \frac{E}{A}x + \frac{F}{A} = 0.$$

If we add to both members

$$\frac{D^2 + E^2}{4A^2},$$

the equation may be placed under the form,

$$\left(y + \frac{D}{2A}\right)^2 + \left(x + \frac{E}{2A}\right)^2 = \frac{D^2 + E^2 - 4AF}{4A^2},$$

which is the equation of a circle, of which the co-ordinates of the centre are $-\frac{D}{2A}$ and $-\frac{E}{2A}$ (Bk. III, Prop. I, Sch. 4), and the radius

$$\frac{\sqrt{D^2+E^2-4AF}}{2A}.$$

The circle may be easily described. For, the centre is determined by laying off from the origin of co-ordinates the distances

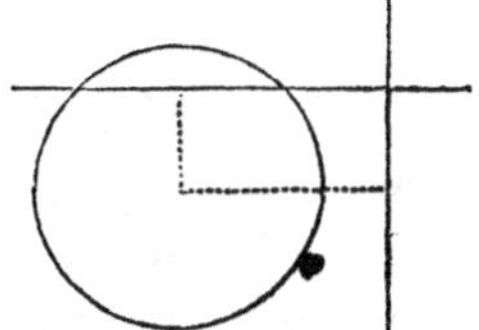

$$-\frac{D}{2A}, \quad -\frac{E}{2A},$$

and drawing parallels to the axes; then knowing the centre and radius, the circumference can be described.

If $D^2+E^2-4AF$ is equal to 0, the circle will reduce to a point; and if it is negative, the circumference will become an imaginary curve.

27. We may conclude from the preceding discussion, that the ellipse, which is characterized by the condition

$$B^2-4AC<0,$$

will become, in succession, the circle, the point, and the imaginary curve, if suitable values and signs be attributed to the constant quantities which enter into the general equation.

## *Of the Parabola.*

$$B^2-4AC=0.$$

28. Let us resume the general value of $y$, which under the present supposition, will be

$$y=-\frac{Bx+D}{2A}\pm\frac{1}{2A}\sqrt{2(BD-2AE)x+D^2-4AF}.$$

If we now make

$$\frac{D^2 - 4AF}{2(BD - 2AE)} = -x',$$

the equation may be put under the form,

$$y = -\frac{Bx + D}{2A} \pm \frac{1}{2A}\sqrt{2(BD - 2AE)(x - x')},$$

in which

$$y = -\frac{Bx + D}{2A}$$

is the equation of the diameter of the curve (Art. 6).

If we suppose $(BD - 2AE)$ to be positive, the curve will extend itself indefinitely in the direction of the positive abscissas, and be limited in the opposite direction.

For, if $x'$ be negative its essential sign under the radical will be positive, and every negative value of $x$ numerically greater than $x'$ will render the factor $x - x'$ negative, and consequently, the corresponding values of $y$ will be imaginary. If $x'$ be positive, as represented in the figure, then, every negative value of $x$, as well as all positive values less than $x'$, will render the values of $y$ imaginary. Under either of the suppositions, of $x'$ negative or $x'$ positive, every positive value of $x$ greater than $x'$ would give real values for $y$: hence, the curve will extend itself indefinitely in the direction of the positive abscissas, and be limited in the opposite direction.

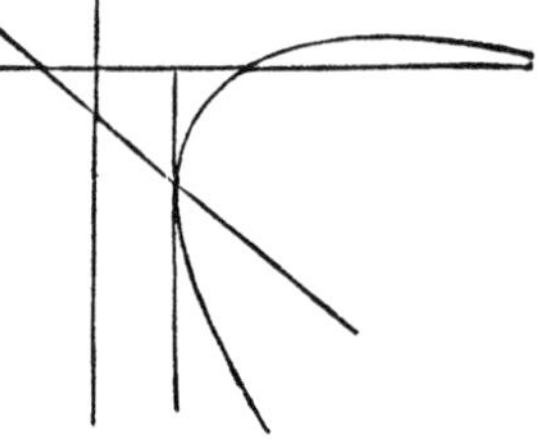

For the value $x = x'$, the radical reduces to 0, and the corresponding value of $y$ is the ordinate of the vertex of the diameter. This ordinate through the vertex is tangent to the curve and limits it in one direction.

If $BD - 2AE$ is negative, then whether $x'$ be positive or negative, every negative value of $x$ greater than $x'$ will render the factor $x - x'$ negative, and consequently give real

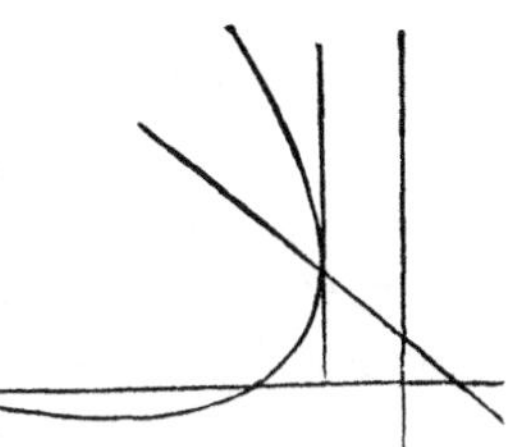

values for $y$. The curve will therefore extend itself indefinitely in the direction of the negative abscissas. It is easily shown that the value $\pm x = \pm x'$ will give the limit of the curve in the other direction.

29. If we resolve the general equation with reference to $x$, he equation of the diameter of the curve will become

$$x = -\frac{(By + E)}{2C};$$

whence, we find

$$y = -\frac{2Cx}{B} - \frac{E}{B}.$$

But the characteristic of the parabola is

$$B^2 - 4AC = 0,$$

whence,

$$\frac{2C}{B} = \frac{B}{2A},$$

and therefore the equation of the second diameter becomes

$$y = -\frac{Bx}{2A} - \frac{E}{B}:$$

hence, this diameter is parallel to the first, of which the equation is

$$y = -\frac{Bx}{2A} - \frac{D}{2A},$$

which proves a property of the parabola already known, viz., that *all diameters of the parabola are parallel to each other* (Bk. V, Prop. VI, Sch. 1).

30. The points at which the parabola intersects the coordinate axes, may be found by combining the equation of the parabola with the equations of the axes: that is, by making $y = 0$ for the axis of $X$, and $x = 0$ for the axis of $Y$.

31. The characteristic of the parabola being

$$B^2 - 4AC = 0,$$

we have

$$B = 2\sqrt{AC};$$

and hence the three first terms,

$$Ay^2 + Bxy + Cx^2,$$

of the general equation, form a perfect square, and may be placed under the form

$$(y\sqrt{A} \pm x\sqrt{C})^2.$$

Observing the directions given in Art. 19 for the discussion of equations, there will be little difficulty in constructing the following:

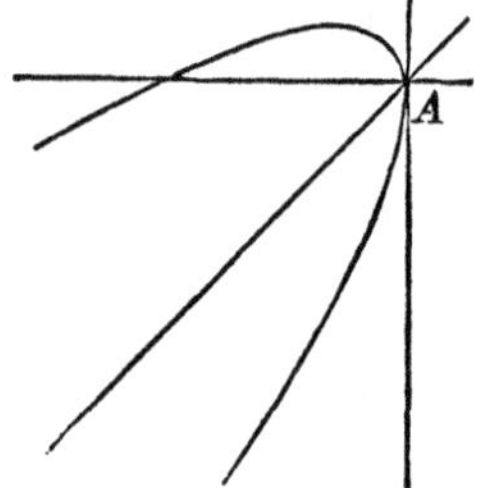


1. $y^2 - 2xy + x^2 + x = 0.$

The equation of the diameter is

$$y = x, \quad \text{or} \quad -y = -x.$$

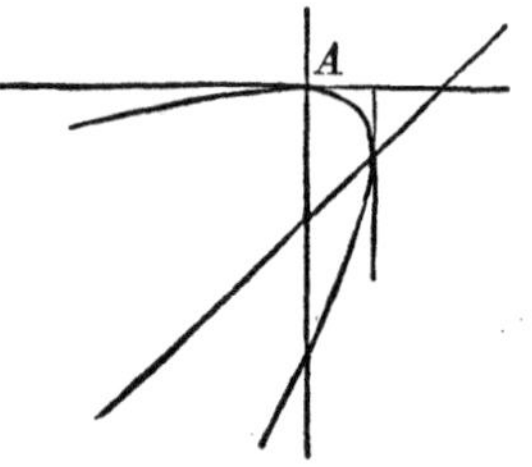


2. $y^2 - 2xy + x^2 + 2y = 0.$

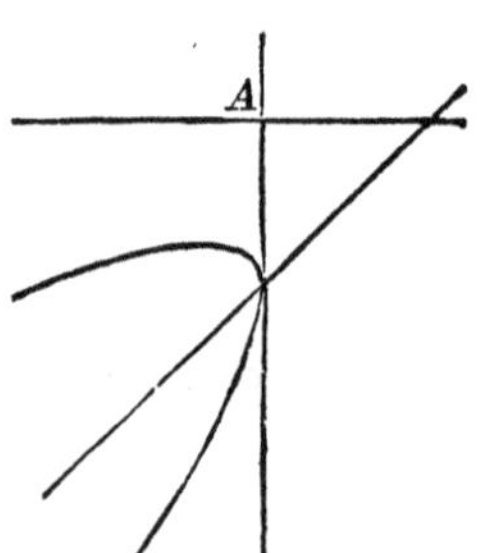


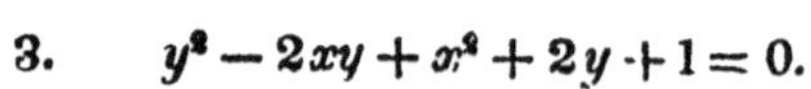

3. $y^2 - 2xy + x^2 + 2y + 1 = 0.$

4. $y^2 - 2xy + x^2 - 2y - 1 = 0.$

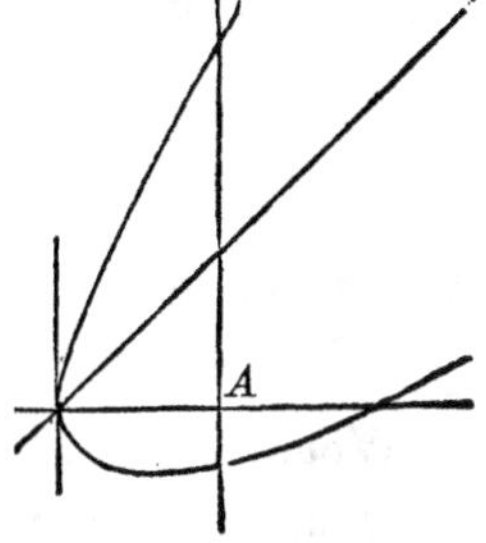


5. $y^2 - 2xy + x^2 - 2y - 2x = 0.$

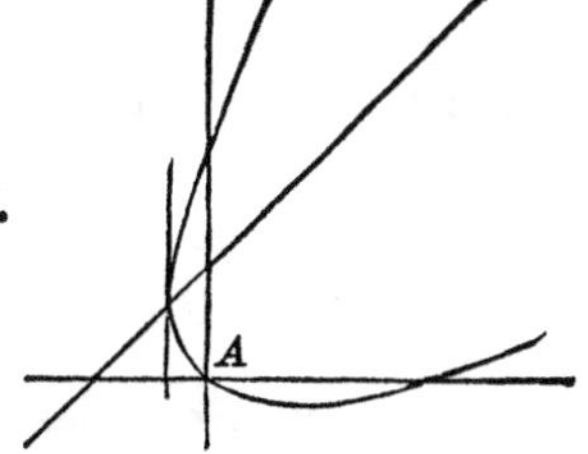


32. If we suppose the factor $(BD - 2AE)$ under the radical, to become 0, the value of $y$ in the general equation will become

$$y = -\frac{Bx + D}{2A} \pm \frac{1}{2A}\sqrt{D^2 - 4AF},$$

or,
$$y = -\frac{Bx}{2A} - \frac{D}{2A} + \frac{1}{2A}\sqrt{D^2 - 4AF}, \qquad \text{and}$$

$$y = -\frac{Bx}{2A} - \frac{D}{2A} - \frac{1}{2A}\sqrt{D^2 - 4AF}.$$

Hence, this supposition reduces the parabola to two straight lines, and since the coefficients of $x$ are the same in the two last equations, the lines will be parallel.

If the quantity $D^2 - 4AF$, under the radical, is positive, the two right lines will be real; if it is 0, they will unite and become the same straight line; and if it is negative the two right lines will be both imaginary.

Under either of these suppositions, the equation

$$y = -\frac{Bx}{2A} - \frac{D}{2A},$$

will always be real, and will represent the diameter as defined in (Bk. V, Prop. VI, Sch. 1).

33. When $B^2 - 4AC = 0$, and $BD - 2AE = 0$, the general equation of the second degree may be resolved into two factors of the first degree : and since the equation will be satisfied by making either of these factors equal to 0, it follows, that the equation ought to represent two right lines. It can be placed under the form,

$$(2Ay + Bx + D + \sqrt{D^2 - 4AF})(2Ay + Bx + D - \sqrt{D^2 - 4AF}),$$

The following examples will illustrate the three cases which have been considered.

1. 

$y^2 - 2xy + x^2 - 1 = 0.$

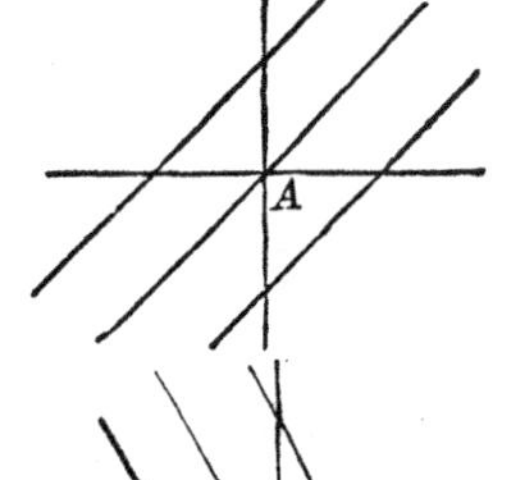


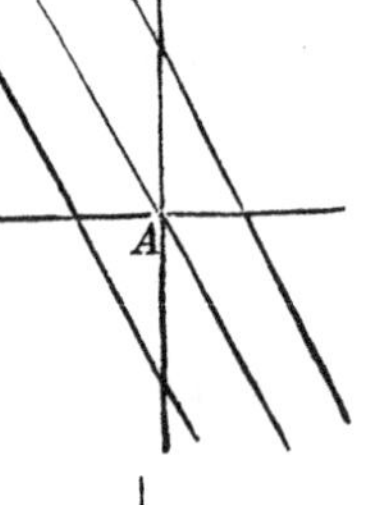


2. $y^2 + 4xy + 4x^2 - 4 = 0.$

3. $y^2 - 2xy + x^2 + 2y - 2x + 1 = 0.$

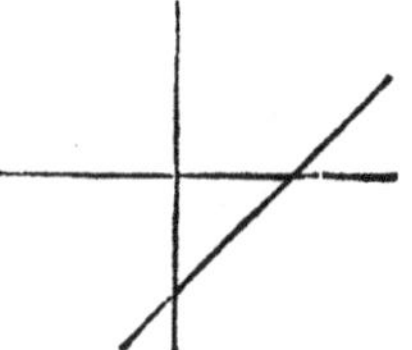

4. $y^2 - 4xy + 4x^2 = 0,$

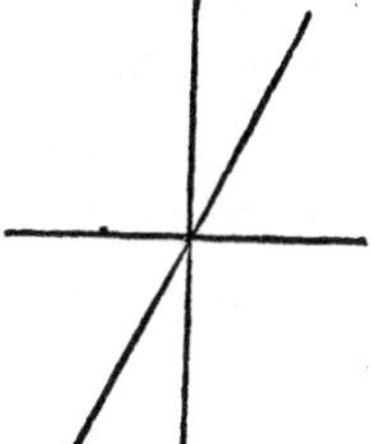

5. $y^2 + 2xy + x^2 + 1 = 0,$

in which $CD$ is the diameter and the parallels imaginary.

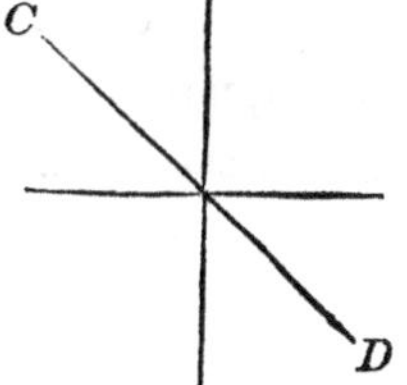


6. $y^2 + y + 1 = 0,$

in which $CD$ is the diameter and the parallels imaginary.

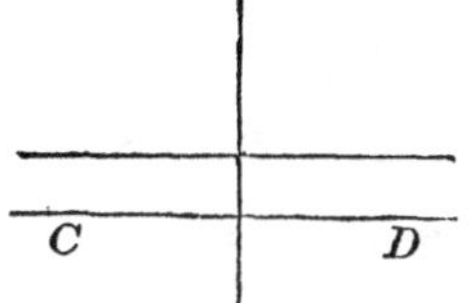


34. We may conclude from the preceding discussion that the parabola which is characterized by the condition

$$B^2 - 4AC = 0,$$

will become, in succession, two parallel straight lines, one straight line, and two imaginary parallels, if suitable values and signs be attributed to the constant quantities which enter into the general equation.

35. Since the diameters of the parabola are infinite, the centre is at an infinite distance from the vertex of the axis, and therefore cannot be used in describing the curve. There is, however, a striking property which was first demonstrated by

Biot, in his Analytical Geometry, and which affords an easy construction for points of the curve. The property is this:

*If through the focus of a parabola any chord be drawn, and two tangents be drawn to the curve at its extremities then,*

1st. *The two tangents will be perpendicular to each other;* and

2d. *Their point of intersection will fall on the directrix.*

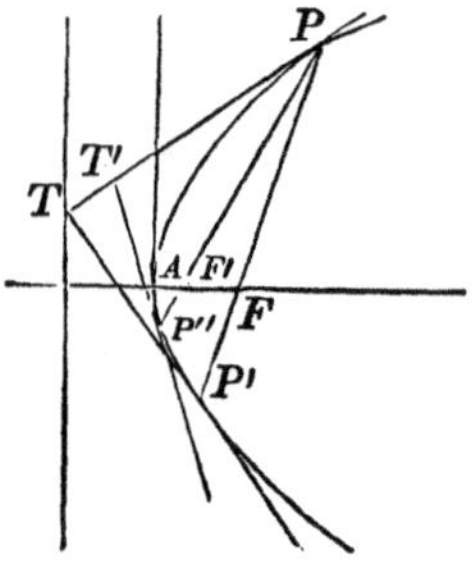

Let there be a parabola of which $A$ is the vertex and $F$ the focus.



Its equation, referred to $A$ as an origin of co-ordinates, is

$$y^2 = 2px.$$

The equation of a straight line passing through a given point, will be of the form

$$y - y' = a(x - x').$$

If the given point be the focus $F$, of which the co-ordinates are $y' = 0$ and $x' = \frac{p}{2}$, we shall have for the equation of the chord,

$$y = a\left(x - \frac{p}{2}\right).$$

If this equation of the chord be combined with that of the parabola, we shall obtain the co-ordinates of the points $P$ and $P'$, in which the chord intersects the curve.

Substituting for $x$ in the equation of the chord its value drawn from the equation of the parabola, and we obtain

$$y^2 - \frac{2p}{a}y - p^2 = 0;$$

in which the two values of $y$ represent the co-ordinates of the points $P$ and $P'$. If we designate these values by $y'$, $y''$, we shall have (Alg. Art. 143),

$$y'y'' = -p^2.$$

If now, through each of the points $P$, $P'$, we conceive a tangent line to be drawn to the parabola, and designate the tangents of the angles which they make with the axis of abscissas by $a'$, $a''$, we shall have (Bk. V, Prop. II),

$$a' = \frac{p}{y'}, \qquad a'' = \frac{p}{y''};$$

and consequently,

$$a'a'' = \frac{p^2}{y'y''}.$$

If we substitute in this result the value of $y'y''$, we have

$$a'a'' = -1, \qquad \text{or} \qquad a'a'' + 1 = 0,$$

and hence, the two tangents are perpendicular to each other.

The converse of this proposition is also true: viz.

*If two tangents are perpendicular to each other, the chord joining the points of tangency will pass through the focus.*

For, let the tangents $PT$, $P'T$, be perpendicular to each other. Draw the chord $PP'$, and let us suppose that it does not pass through the focus. From $P$, draw the chord $PF'P''$ through the focus $F'$, and through $P''$ draw a tangent to the curve: then by what has just been demonstrated it will be perpendicular to $PT$: hence, the tangents $P'T$ and $P''T'$ will be parallel, since they are both perpendicular to $PT$. But the tangent of the angle which a tangent line makes with the axis of abscissas is

$$\frac{p}{y''};$$

$y''$ being the ordinate of the point of contact.

Now, since no two points of the parabola have equal ordinates with the same sign, it follows that two tangents drawn at different points cannot be parallel: hence, $P''$ must coincide with $P'$, and $P''T'$ with $P'T$: therefore, the chord which joins the points of tangency will pass through the focus

36. We will now demonstrate the second part of the proposition, viz: that the point of intersection of the tangents falls on the directrix.

For this purpose let us designate the co-ordinates of the points of tangency by $x'$, $y'$ and $x''$, $y''$: the equations of the tangents will then become

$$yy' = p(x + x'), \qquad yy'' = p(x + x'').$$

For the common point, or point of intersection, we shall have $y = y$ and $x = x$: hence, if we combine the two equations by dividing them member by member, we shall have

$$\frac{y''}{y'} = \frac{(x + x'')}{(x + x')},$$

from which we find

$$x = \frac{x''y' - x'y''}{y'' - y'},$$

in which $x$ represents the abscissa of the point of intersection of the tangents. This value may be simplified by recollecting that the points of contact are on the curve, and hence,

$$y'^2 = 2px', \qquad y''^2 = 2px''.$$

Finding the values of $x'$, $x''$, and substituting them in the last equation, and then dividing by the common factor $y'' - y'$, we find

$$x = \frac{y'y''}{2p}.$$

As no condition has yet been introduced fixing the position of the chord or the direction of the tangents, it follows, that this is a general expression for the abscissa of the point in which two tangents intersect each other when drawn through the extremities of any chord.

If now, we take the chord passing through the focus, we have just found that

$$y'y'' = -p^2.$$

Introducing this condition into the last equation, and we have

$$x=-\frac{p}{2},$$

that is, the abscissa of the point of intersection is without the curve, and at a distance from the origin equal to $-\frac{p}{2}$ which is the distance of the directrix from the same point. Hence, the point of intersection of the tangents, falls on the directrix.

37. To apply this property in the construction of the curve, it should be recollected that when the general equation is resolved with reference to $y$, we obtain a diameter of which the ordinates and the tangent line at the vertex are parallel to the axis of $Y$: and if we resolve the equation with reference to $x$, we shall obtain a diameter of which the ordinates and the tangent line at the vertex are parallel to the axis of $X$. Now since the co-ordinate axes are at right angles to each other, these tangents will be at right angles; and hence, the chord joining the points of tangency will pass through the focus.

These points of tangency are easily determined; for they are the limits of the curve in the directions of the axes. For the tangent parallel to the axis of $Y$, we have already found (Art. 28),

$$x'=-\frac{D^2-4AF}{2(BD-2AE)},$$

which gives, - $y'=-\frac{Bx'+D}{2A}$.

If we now resolve the equation with reference to $x$, we shall find the co-ordinates of the vertex of the other diameter

At this vertex, the tangent is parallel to the axis of $X$. We thus find

$$y'' = -\frac{E^2 - 4CF}{2(BE - 2CD)},$$

which gives $x'' = -\frac{By'' + E}{2C}$.

Having thus found the two points of tangency $P$, $P'$, join them by a straight line $PP'$: this line will pass through the focus.

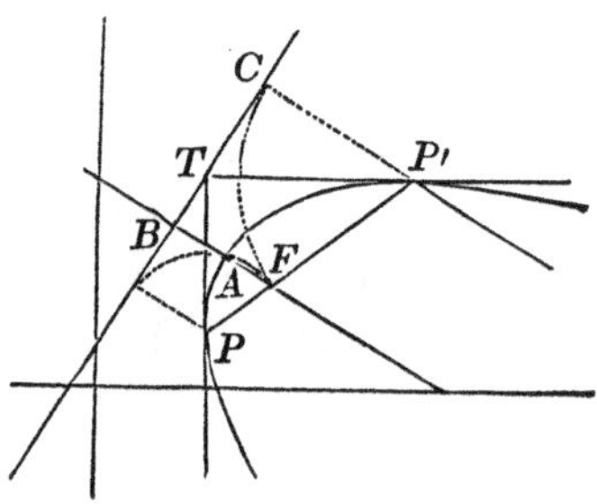


Through the point $P$, draw $PT$ parallel to the axis of ordinates, and through $P'$ draw $P'T$ parallel to the axis of abscissas: these lines will be tangent to the parabola, and perpendicular to each other: and hence, their point of intersection $T$, will be on the directrix.

Having constructed either of the diameters, the one, for example through $P$, draw through $T$, $TB$ perpendicular to it, and $TB$ will be the directrix of the parabola.

With $P'$ as a centre, and a radius equal to $PC$, the distance to the directrix, describe the arc $CF$: the point $F$, at which it intersects $PP'$, will be the focus. The line $FB$, drawn through $F$ and perpendicular to the directrix, will be the axis; and the point $A$, equally distant from $F$ and $B$, will be the vertex.

## *Of the Hyperbola*

$$B^2 - 4AC > 0.$$

38. After what has preceded, there can be little difficulty in discussing this class of curves

Let us resume the general value of $y$, and place it under the form

$$y=-\frac{Bx+D}{2A}\pm\frac{1}{2A}\sqrt{(B^2-4AC)\left\{x^2+2\frac{BD-2AE}{B^2-4AC}x+\frac{D^2-4AF}{B^2-4AC}\right\}}$$

If we represent by $x'$, $x''$, the two roots of the equation

$$x^2+2\frac{BD-2AE}{B^2-4AC}x+\frac{D^2-4AF}{B^2-4AC}=0,$$

the values of $y$ may be put under the form

$$y=-\frac{Bx+D}{2A}\pm\frac{1}{2A}\sqrt{(B^2-4AC)(x-x')(x-x'')}.$$

There are here, as in the ellipse, three cases:

1st. When the roots $x'$, $x''$, are real and unequal, of which the condition is

$$(BD-2AE)^2-(B^2-4AC)(D^2-4AF)>0.$$

2d. When they are equal, of which the condition is

$$(BD-2AE)^2-(B^2-4AC)(D^2-4AF)=0.$$

3d. When they are both imaginary, of which the condition is

$$(BD-2AE)^2-(B^2-4AC)(D^2-4AF)<0.$$

1st. *When the roots are real and unequal.*

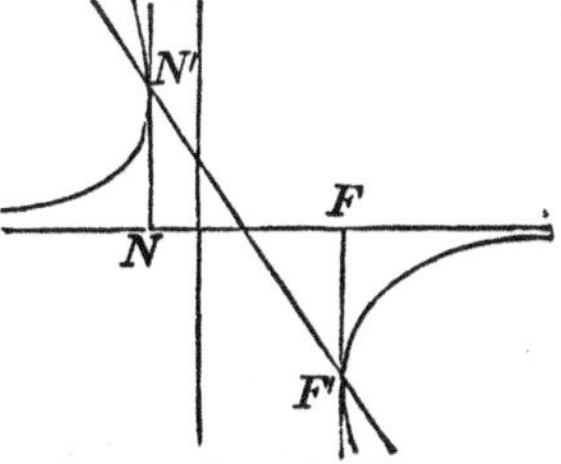


39. When the roots are real and unequal, all values of $x$ greater than $x'$ and less than $x''$, will give contrary signs to the factors $x-x'$, $x-x''$: their product will therefore be negative, and as $B^2-4AC$ is positive, the quantity under the radical sign will be negative: hence, the curve will be imaginary between the limits $x'$ and $x''$, and will extend indefinitely beyond these limits.

If the abscissas $x'$, $x''$, be substituted in the equation of the diameter

$$y = -\frac{Bx + D}{2A},$$

we shall find the ordinates of the vertices $F'$, $N'$: they are,

$$y' = -\frac{Bx' + D}{2A}, \quad \text{and} \quad y'' = -\frac{Bx'' + D}{2A}.$$

The points in which the curve intersects the co-ordinate axes, may be found by combining the equation of the curve with the equations of the axes.

The following examples will illustrate what has been explained:

1. $y^2 - 2xy - x^2 + 2 = 0.$

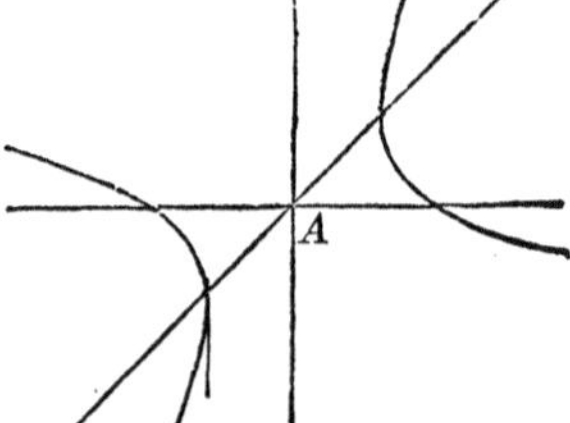


2. $y^2 - x^2 + 2x - 2y + 1 = 0.$

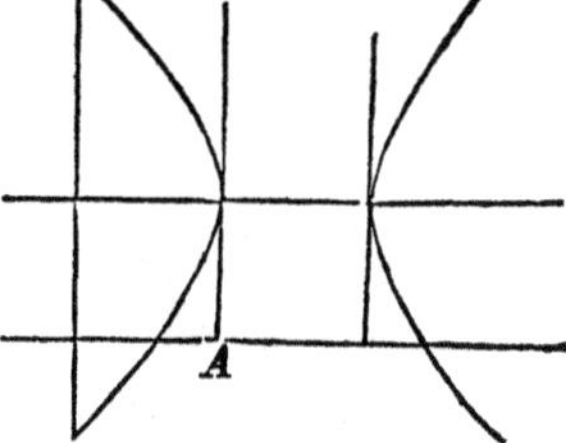


3. $y^2 - 2xy - x^2 - 2y + 2x + 3 = 0.$

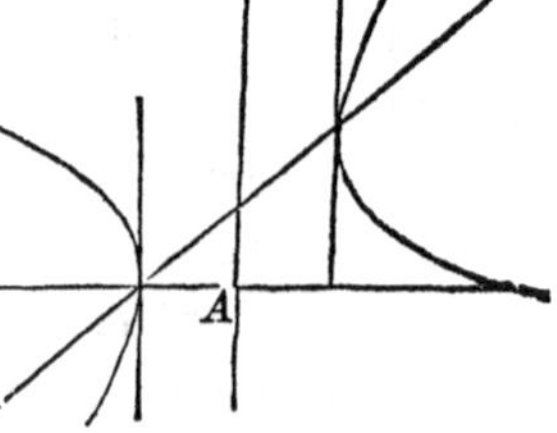

4. $y^2-2x^2-2y+6x-3=0.$

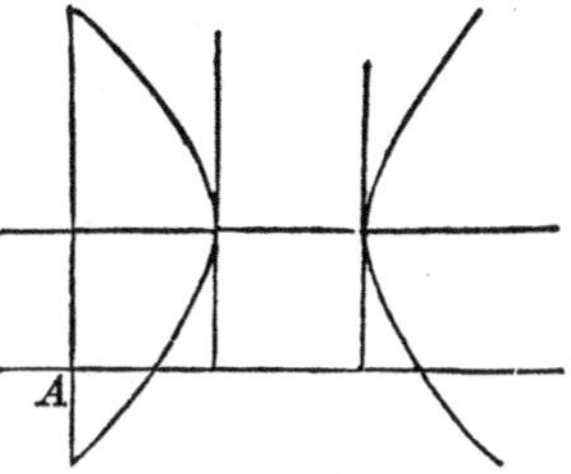


2d. *When the roots* x′, x″, *are equal.*

40. When the roots $x'$, $x''$, become equal to each other, the value of $y$ becomes

$$y=-\frac{Bx+D}{2A}\pm\frac{x-x'}{2A}\sqrt{B^2-4AC};$$

and this value is always real, since the quantity under the radical is positive.

This equation will represent two right lines, and since the coefficients of $x$ will be different in each, the lines will intersect each other. The line $AB$ of which the equation is

$$y=-\frac{Bx+D}{2A},$$

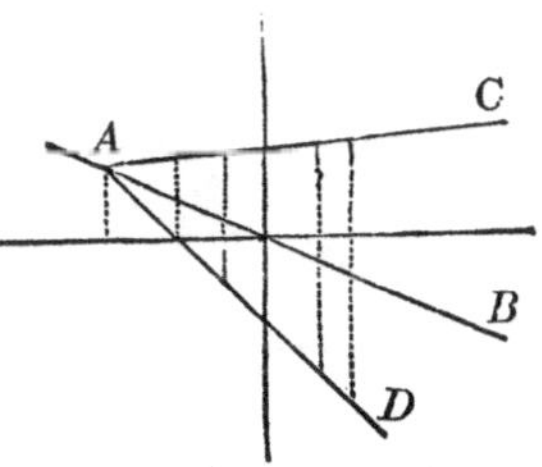


will bisect all lines that are limited by the two straight lines, $AD$, $AC$, and parallel to the axis of $Y$. Hence, it may be still considered as a diameter.

If we make $x=x'$, we shall have

$$y'=-\frac{Bx'+D}{2A}.$$

which gives the ordinate of the point $A$, where the two lines intersect each other.

The condition $x' = x''$ enables us to place the general equation under the form

$$\left\{2Ay + Bx + D - (x - x')\sqrt{(B^2 - 4AC)}\right\} \times$$

$$\left\{2Ay + Bx + D + (x - x')\sqrt{(B^2 - 4AC)}\right\} = 0.$$

This equation is composed of two factors of the first degree, and can therefore be satisfied by making either of them equal to 0. It ought therefore to be the equation of two straight lines.

We shall add a few examples.

1. $y^2 - 2x^2 + 2y + 1 = 0.$

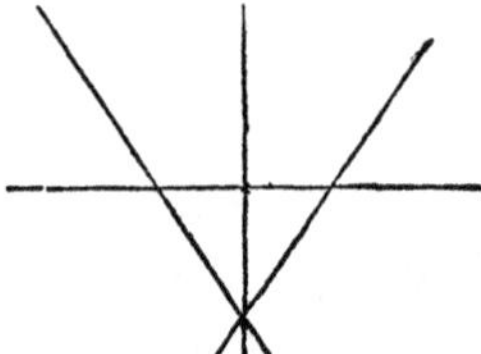

2. $y^2 - x^2 = 0.$

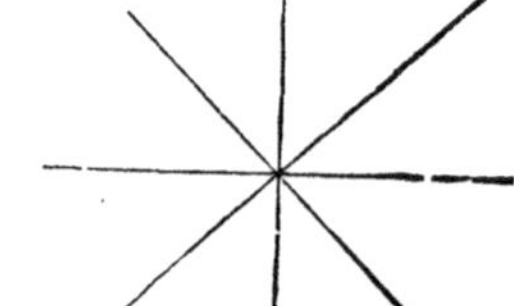

3 $y^2 + xy - 2x^2 + 3x - 1 = 0$

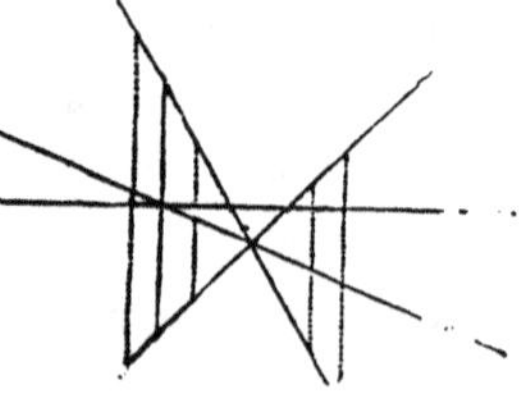

3d. *When the roots* x′, x″, *are imaginary.*

**41.** When in the equation

$$y = -\frac{Bx + D}{2A} \pm \frac{1}{2A}\sqrt{(B^2 - 4AC)(x - x')(x - x'')}$$

the roots $x'$, $x''$, are imaginary, the product of the factors $(x-x')$, $(x-x'')$, will be always positive (Art. 23), and since $B^2-4AC$ is positive, the quantity under the radical will be positive, and hence $y$ will be real for all values of $x$, and each abscissa will give two values for $y$.

The part which is independent of the radical

$$y=-\frac{2Bx+D}{2A}.$$

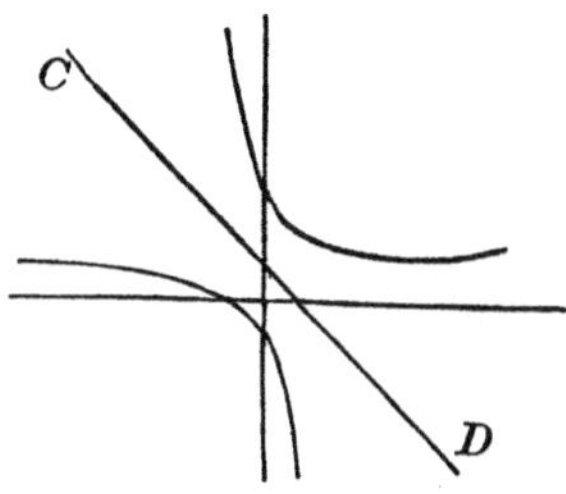


will still be the equation of a diameter of the curve, and this diameter $CD$ will bisect all lines parallel to the axis of $Y$ and terminated by the curve. The line $CD$, is therefore a diameter of the hyperbola.

The imaginary values $x'$, $x''$, are the roots of the equation which was obtained by placing the quantity under the general radical equal to 0. The values of $x$ which would reduce that quantity to 0, correspond to the vertices of the diameter. But as there are no values of $x$ which will satisfy the equation, it follows that the diameter does not intersect the curve.

Let the hyperbolas of which the following are the equations be constructed.

1. $y^2-2xy-x^2-2=0.$

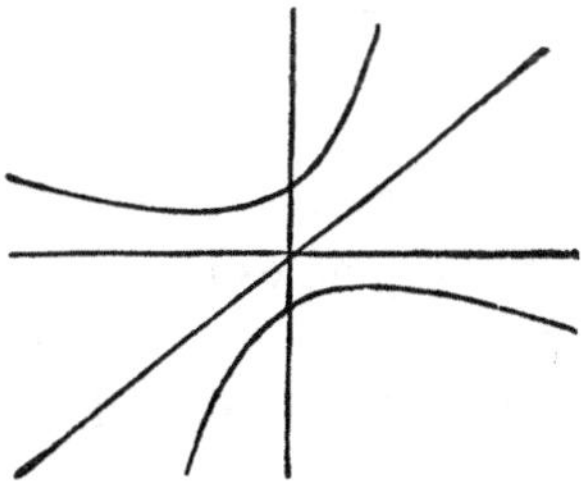

2. $y^2+2xy-x^2+2x+2y-1=0.$

3. $y^2-2xy-x^2-2x-2=0.$

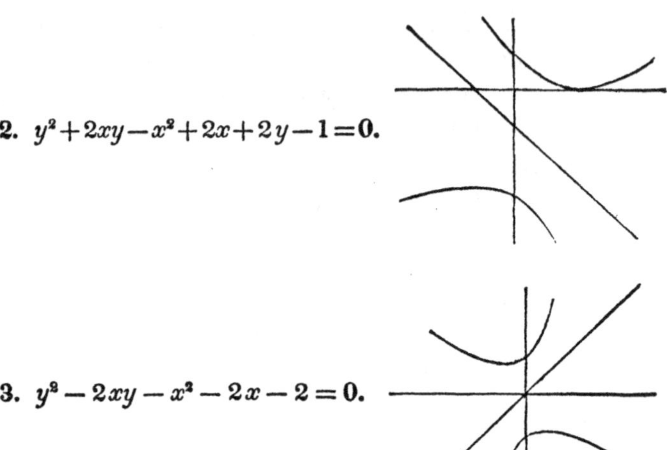

42. Having found the abscissas $x'$, $x''$ of the vertices of that diameter which intersects the curve (Art. 37), we can readily find the value of this diameter by a method entirely similar to that pursued in (Art. 15).

The length of the diameter will be equal to

$$\frac{(x''-x')}{2A}\sqrt{B^2+4A^2},$$

and the length of its conjugate diameter to

$$\frac{x''-x'}{2A}\sqrt{4AC-B^2}.$$

The conjugate diameter will be imaginary, since $4AC-B^2$, is negative. If we designate the angle included by the diameters by $\alpha$, we shall have, as in the ellipse (Art. 16),

$$\sin\alpha=\frac{2A}{\sqrt{4A^2+B^2}}.$$

Then, knowing the conjugate diameters and the angle

which they form with each other, the hyperbola may be described.

43. There is yet one case to be examined: it is that in which

$$A = -C \quad \text{and} \quad B = 0.$$

Under this supposition the general equation will become

$$A\,y^2 - Ax^2 + Dy + Ex + F = 0,$$

or, $$y^2 - x^2 + \frac{D}{A}y + \frac{E}{A}x + \frac{F}{A} = 0:$$

and, by adding $\frac{D^2}{4A^2}$ and $-\frac{E^2}{4A^2}$, to both members, it may be put under the form

$$\left(y + \frac{D}{2A}\right)^2 - \left(x - \frac{E}{2A}\right)^2 = \frac{D^2 - E^2 - 4AF}{4A^2},$$

which represents an equilateral hyperbola of which the co-ordinates of the centre $C$ are

$$-\frac{D}{2A} \quad \text{and} \quad +\frac{E}{2A},$$

and of which the power is one half of

$$\frac{D^2 - E^2 - 4AF}{4A^2}.$$

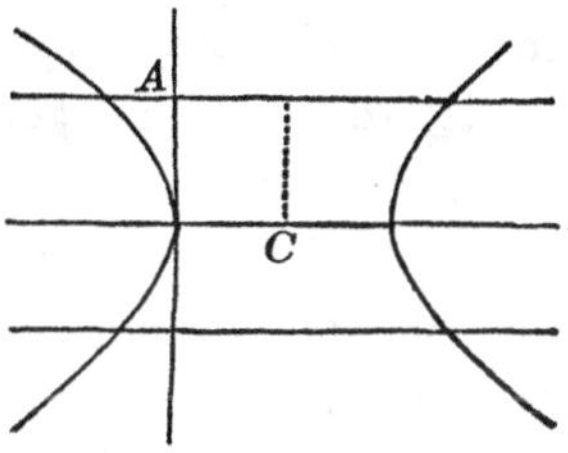

This case corresponds to that in which the ellipse becomes the circle.

## *Of the Centres and Diameters of Curves.*

44. That point in the plane of a curve which bisects all straight lines drawn through it and limited by the curve, is called the *centre*.

Admitting, for a moment, that there is such a point, let us

suppose the origin of co-ordinates to be transferred to it. Through the origin let any straight line be drawn, and designate the co-ordinates of one of the points in which it intersects the curve by $+x'$, $+y'$; the co-ordinates of the point in which it intersects the curve, on the other side of the centre, will be designated by $-x'$, $-y'$. Now, since this is true for every position of the right line, it follows, that if the curve has a point whose co-ordinates are $+x'$, $+y'$, it must also have another point whose co-ordinates are $-x'$, $-y'$. But, if an equation is equally satisfied for all values of $+x'$, $+y'$, and $-x'$, $-y'$, it must necessarily be of such a form as to undergo no alteration when the signs of both the variables are changed from $+$ to $-$, or from $-$ to $+$.

This condition requires, that every term of the equation should be of an even degree. If, therefore, a curve have a centre, all the terms of its equation, when referred to it as an origin of co-ordinates, will be of an even degree.

To determine then whether a curve has a centre, it is simply necessary to inquire whether its equation can be reduced to such a form that each term shall be of an even degree with respect to the variables.

For this purpose, let us assume the formulas for transferring the origin of co-ordinates, without changing the direction of the axes. They are

$$x = a + x', \qquad y = b + y'$$

If we now substitute these values of $x$ and $y$, in the general equation of the second degree,

$$Ay^2 + Bxy + Cx^2 + Dy + Ex + F = 0,$$

we shall have

$$\left.\begin{array}{c} Ay'^2 + Bx'y' + Cx'^2 + Ab^2 + Bab + Ca^2 + Db + Ea + F \\ (2Ab + Ba + D)y' + (2Ca + Bb + E)x' \end{array}\right\} = 0$$

Since the co-ordinates $a$ and $b$ of the new origin are

entirely arbitrary, we can in general attribute to them such values as will introduce two conditions into the last equation. We are therefore at liberty to make

$$2Ab+Ba+D=0, \quad \text{and} \quad 2Ca+Bb+E=0.$$

If, in addition, we represent the known terms of the las equation by $P$, the equation will take the form

$$Ay'^2+Bx'y'+Cx'^2+P=0,$$

an equation which will not be affected by changing $+x', +y'$ into $-x', -y'$, or $-x', -y'$ into $+x', +y'$.

45. The existence of a centre depends entirely on the two equations,

$$2Ab+Ba+D=0, \quad \text{and} \quad 2Ca+Bb+E=0.$$

If we suppose $a$ and $b$ to assume all possible values, which will satisfy the two equations, each will represent a straight line, and they may be placed under the forms

$$b=-\frac{B}{2A}a-\frac{D}{2A}, \qquad b=-\frac{2C}{B}a-\frac{E}{B}.$$

As the centre must lie on each of these right lines, it will be found at their intersection. Hence, if we combine the equations, the values of $a$ and $b$ will be the co-ordinates of the centre. They are

$$a=\frac{2AE-BD}{B^2-4AC}, \qquad b=\frac{2CD-BE}{B^2-4AC}.$$

These values will be single and finite so long as $B^2-4AC$ is not 0: hence, there is always one point, and but one, in each of the planes of the ellipse and hyperbola which enjoys the properties of a centre.

When $B^2-4AC=0$, which is the characteristic of the parabola, the values of both $a$ and $b$ become infinite, which shows that the centre is at an infinite distance from both axes, or in other words, that the parabola has no centre.

From the characteristic of the parabola $B^2-4AC=0$, we find

$$\frac{B}{2A}=\frac{2C}{B};$$

from which we see, that the two lines, which by their intersection determine the centre of the curve, become parallel to each other: hence, the co-ordinates of the centre ought to be infinite, or the centre ought not to exist.

46. It has been shown, in Art. 32, that when

$$(BD-2AE)=0,$$

the parabola reduces to two straight lines.

This supposition also gives

$$\frac{D}{2A}=\frac{E}{B};$$

and hence, the two lines which determine the centre will coincide with each other: the centre will therefore be at any point of the common line.

This indetermination arises from the relation which exists between the coefficients of $a$ and $b$, in the equations

$$2Ab+Ba+D=0, \qquad 2Ca+Bb+E=0;$$

that is, if one of the equations is equal to 0, the relation between the coefficients is such as to satisfy the other: and hence the conditions are not independent; and therefore the two equations ought not to determine $a$ and $b$ (Alg. Art 103).

It should, however, be remarked, that when this indetermination arises, the parabola is reduced to two parallel straight lines, and the centre is limited to a straight line parallel to, and equally distant from them; and this is equally true, whether the parallels are real or imaginary.

**47.** It has already been remarked (Bk. V, Prop. VI, Sch. 1), that the term *diameter*, designates any straight line which bisects a system of chords drawn parallel to each other and limited by the curve.

If, therefore, the axis of abscissas coincides with a diameter of the curve, and the axis of ordinates be drawn parallel to the system of chords which the diameter bisects, the curve will be symmetrical with respect to the axis of abscissas; for, each value of $x$ will give two equal values of $y$ with contrary signs.

Under this supposition, therefore, the equation of the curve will not contain the first power of $y$.

Again, if we suppose the axis of $Y$ to coincide with a diameter of the curve, and the axis of $X$ to be parallel to the system of chords which are bisected by the diameter, each value of $y$ will give two equal values of $x$ with contrary signs: hence, under this supposition, the equation will not contain the first power of $x$.

If, therefore, it be required to determine whether the curves represented by the general equation of the second degree,

$$Ay^2 + Bxy + Cx^2 + Dy + Ex + F = 0,$$

have one or more diameters, it will only be necessary to ascertain, whether the equation can be so transformed as to cause the terms containing the first power of one or both of the variables to disappear.

For this purpose, let us refer the curve to a new system of co-ordinates. Since the axes of the primitive system have been supposed at right angles to each other, we shall require the formulas

$$x = a + x' \cos \alpha + y' \cos \alpha', \quad y = b + x' \sin \alpha + y' \sin \alpha',$$

for passing from a system of rectangular to a system of oblique axes.

Substituting these values of $x$ and $y$ in the general equation, and we obtain,

$[2A\sin\alpha\sin\alpha' + B(\sin\alpha\cos\alpha' + \sin\alpha'\cos\alpha) + 2C\cos\alpha\cos\alpha']x'y'$,
$+[(2Ab + Ba + D)\sin\alpha + (2Ca + Bb + E)\cos\alpha]x'$,
$+[(2Ab + Ba + D)\sin\alpha' + (2Ca + Bb + E)\cos\alpha']y'$,
for the terms which contain the uneven powers of $x'$ and $y'$

By referring the curve to a new system of co-ordinate axes, we have introduced into its equation four arbitrary constants, $a$, $b$, $\alpha$, and $\alpha'$: we can therefore attribute to these constants such values as will introduce four independent conditions into the resulting equation.

If then, we place, as we are at liberty to do, the coefficients of the first powers of $y'$ equal to 0, the new axis $X'$ will become a diameter of the curve. The analytical conditions are,

$$2C\cos\alpha\cos\alpha' + B(\sin\alpha\cos\alpha' + \sin\alpha'\cos\alpha) + 2A\sin\alpha\sin\alpha' = 0,$$

which may be placed under the form

$$2C + B(\tan\alpha' + \tan\alpha) + 2A\tan\alpha\tan\alpha' = 0; \quad (1)$$

and

$$(2Ab + Ba + D)\sin\alpha' + (2Ca + Bb + E)\cos\alpha' = 0. \quad (2)$$

In order to render the new axis $Y'$ a diameter, we must place the coefficients of $x'$ equal to 0: which gives,

$$2C + B(\tan\alpha' + \tan\alpha) + 2A\tan\alpha\tan\alpha' = 0, \quad (1)$$

and

$$(2Ab + Ba + D)\sin\alpha + (2Ca + Bb + E)\cos\alpha = 0. \quad (3)$$

The first equation in each of the conditions is the same, as indeed it ought to be, since the coefficient of the product $x'y'$ is alike the coefficient of the first power of $x'$, or of $y'$.

48. Let us now examine more particularly equations (1), (2) and (3).

In order that the new axis of $X'$ shall be a diameter, that is, in order that it shall bisect a system of chords parallel to the axis of $Y'$, equations (1) and (2) must be satisfied at the same time. Since these two equations contain the four arbitrary constants $a$, $b$, $\alpha$ and $\alpha'$, we might suppose ourselves a.

liberty to assign arbitrary values to two of them, and then satisfy the equations by attributing suitable values to the other two. But since neither $a$ nor $b$ enters into equation (1), it is evident that we cannot attribute arbitrary values to $\alpha$ and $\alpha'$, for equation (1) will always determine one of them when the other is known. We may, however, if we please attribute arbitrary values to $a$, and $b$, and this will fix the position of the new origin. Under this supposition equation (2) will determine the value of $\alpha'$, and consequently the direction of the new axis $Y'$ will be known. Substituting the value of $\alpha'$ in equation (1) we find the value of $\alpha$, and hence the direction of the new axis of $X'$ becomes known.

Under these suppositions equation (3) will not in general be satisfied, since all the arbitrary constants which enter into it, have been determined by other conditions.

If we attribute a given value to $\alpha$ or $\alpha'$, that is, if we assume the direction of one of the new axes, the direction of the other will be determined by equation (1), and consequently, the angle formed by the new axes will become known.

If then we wish that both of the new axes shall be diameters, we must satisfy at the same time equations (2) and (3) by attributing suitable values to $a$ and $b$. But since $a$ and $b$ do not depend on $\alpha$ and $\alpha'$, and since the equations are to be satisfied for all values of $\alpha$ and $\alpha'$ we have (Alg. Art. 208),

$$2\,Aa + Bb + D = 0 \quad \text{and} \quad 2\,Ca + Bb + E = 0.$$

But these are the equations of the centre of the curve (Art. 45); and consequently the co-ordinates of the centre will satisfy the equation of every diameter. Hence, every diameter of a curve of the second degree passes through the centre. Reciprocally, every line passing through the centre is a diameter, since if we satisfy equation (2) or (3) independently of $\alpha$ or $\alpha'$ we can always satisfy equation (1), and hence the conditions of a diameter will be fulfilled.

49. If it be required that the first power of neither variable shall appear in the transformed equation, the three conditions,

$$2\,C + B(\text{tang }\alpha' + \text{tang }\alpha) + 2\,A \text{ tang }\alpha \text{ tang }\alpha' = 0, \quad (1)$$

$$2\,Ab + Ba + D = 0, \quad (2) \qquad 2\,Ca + Bb + E = 0, \quad (3)$$

must subsist at the same time, and the transformed equation will take the form

$$A'y'^2 \pm B'x'^2 \mp P = 0\,;$$

and since each axis will then bisect all chords which are parallel to the other, the curve will be referred to its centre and conjugate diameters.

Since there are still two undetermined quantities, $\alpha$ and $\alpha'$, in equation (1), it follows, that there is an infinite number of systems of conjugate diameters which will fulfil the conditions.

If, however, a given value be attributed either to $\alpha$ or $\alpha'$ the value of the other may be determined from equation (1); and if we make $\alpha' - \alpha = 90$, the curve will be referred to its centre and axes.

50. The last equation has been obtained under the supposition that equations (2) and (3) may be both satisfied at the same time by finite values of $a$ and $b$. We have seen, however (Art. 45), that when the curve becomes a parabola, the conditions which will satisfy one of the equations will cause the two straight lines which they represent to become parallel to each other. The parabola will then be symmetrical with respect to either of the parallels determined by equations (2) and (3).

51. In reviewing the methods which have been pursued in the discussion of the general equation of the second degree, we see, that the equation has been simplified by transferring the origin of co-ordinates, and changing the directions of the axes.

In the ellipse and hyperbola, which are characterized by

$$B^2 - 4\,AC < 0, \qquad \text{and} \qquad B^2 - 4\,AC > 0,$$

the first power of the variables can be made to disappear by simply changing the position of the origin.

The equation which establishes such a relation between $\alpha'$ and $\alpha$ that one of the new axes shall bisect all chords of the curve which are parallel to the other, will render the coefficient of the rectangle of the variables equal to 0, and will therefore cause that term to disappear from the transformed equation. The last condition has no reference to the position of the origin of co-ordinates: it merely expresses a relation between the angles which the new axes form with the primitive axis of abscissas.

52. The condition $B^2 - 4AC = 0$, which characterizes the parabola, attributes such values to the constants which enter into the coefficients of $x'$, $y'$, in the transformed equation, that by placing one of them equal to 0, the other cannot be reduced to 0 for finite values of $a$ and $b$ (Art. 45). In this case, the two conditions expressed by equations (2) and (3) are dependant on each other. If either of the new axes $X'$ or $Y'$ is a diameter, we have $B = 0$. Now, if $X'$ is a diameter, we have in addition to the last condition $D = 0$. Under this supposition $A$ cannot be 0, for then the general equation would not contain $y$. Hence the condition $B^2 - 4AC = 0$, would give $-4 AC = 0$ and consequently $C = 0$ or the second power of $x$ would not appear in the equation.

If $Y'$ is a diameter, we have $B = 0$ and $E = 0$. then $C$ cannot be 0; hence $A = 0$, or the second power of $y$ will not enter the equation. Hence the equation of the parabola must reduce to the forms.

$$y'^2 = Px' + \alpha,$$

or

$$x'^2 = P'y' + \alpha',$$

or by again transferring the origin of co-ordinates to

$$y'^2 = Px'',$$

$$x'^2 = P'y''.$$

# BOOK VIII.

## *Of the Point and Straight Line in Space—Of the Plane—Of the Transformation of Co-ordinates in Space—Of Polar Equation in Space.*

1. Space is indefinite extension and is entirely similar in all its parts. We are therefore unable to determine the *absolute* places or positions of the geometrical magnitudes which are to be subjected to the algebraic analysis, since there is nothing fixed to which they can be referred. Their *relative* positions may, however, be easily determined and these will enable us to discuss and develop their properties.

2. Thus far, the analysis has been limited to points and lines lying in the same plane, and these have been referred to two straight lines making a given angle with each other. The analysis is now to be extended to points and lines in space, and these will be referred to three planes, which for simplicity, will be taken at right angles to each other.

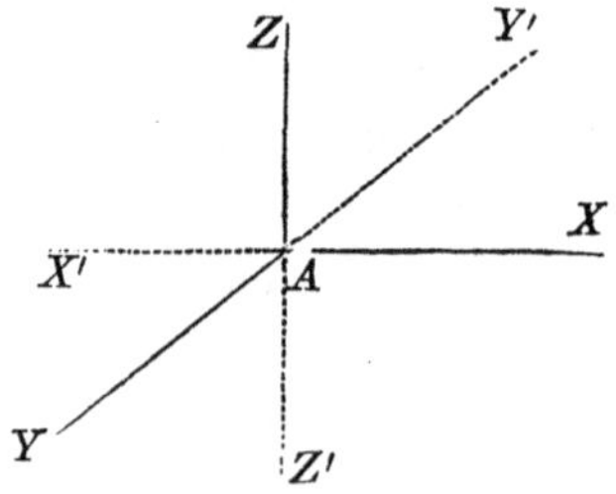

3. Let $AX$, $AY$, $AZ$, designate the three straight lines in which the planes intersect each other. The plane $ZAX$ is supposed to be vertical and to coincide with the plane of the paper. The plane $YAX$ is supposed to be horizontal, and to intersect $ZAX$ in the horizontal line $X'X$. The plane $YAZ$ is perpendicular to the other two planes and intersects the horizontal plane in the horizontal line $YY'$, and the vertical plane in the vertical line $ZZ'$. The three planes are called, the *co-ordinate planes*.

4. Since the co-ordinate planes are respectively at right angles to each other, the line of intersection of either two will be perpendicular to the third: and this line of intersection is called the *axis* of that plane to which it is perpendicular.

For example, $X$ is the axis of the co-ordinate plane $YZ$, $Y$ the axis of the co-ordinate plane $ZX$, and $Z$ the axis of the co-ordinate plane $YX$. The three are called, the *co-ordinate axes*, and their point of intersection $A$, is called the *origin*.

5. The co-ordinate planes are supposed to be indefinite, and hence, they will divide all space into eight equal parts, or solid angles, having the origin $A$ for a common vertex. Four of these angles are above the horizontal plane $YAX$, and four below it. They are thus designated.

| | | |
|---|---|---|
| $YAX$ | is called the | 1st angle, |
| $YAX'$ | " " | 2d " |
| $X'AY'$ | " " | 3d " |
| $Y'AX$ | " " | 4th " |

The fifth angle is directly beneath the first, the sixth beneath the second, the seventh beneath the third, and the eighth beneath the fourth.

This manner of naming the angles differs from that adopted in the plane, where the first angle is beyond the axis of abscissas, and where we pass round from the right to the left; but both the methods are now too well established to be changed merely for the purpose of producing uniformity.

6. The distance of any point in space from either of the co-ordinate planes is estimated on the axis of the plane, or on a line parallel to the axis.

7. Let us suppose that we know the distances of a point from the three co-ordinate planes, viz:

from $YZ = a$,

from $ZX = b$,

from $YX = c$.

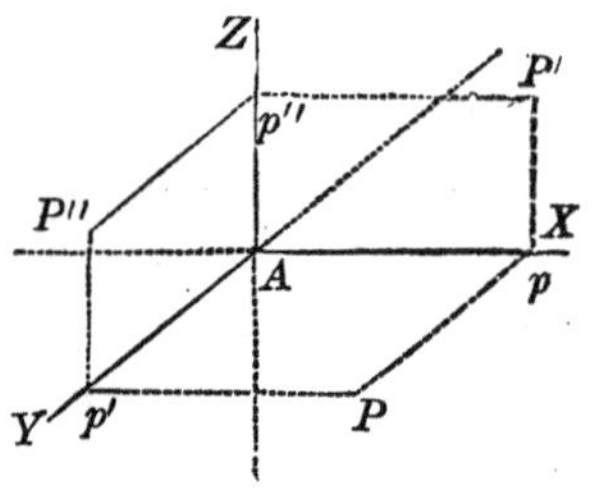

From the origin $A$ lay off on the axis of $X$, a distance $Ap = a$, and through $p$ pass a plane parallel to the co-ordinate plane $YZ$. Its traces $pP'$, $pP$, will be respectively parallel to the axes $Z$ and $Y$. Lay off in like manner on the axis of $Y$, a distance $Ap' = b$, and through $p'$ pass a plane parallel to the co-ordinate plane $ZX$. Its traces $p'P$, $p'P''$ will be respectively parallel to the axes $X$ and $Z$. Since the point must be in both planes at the same time, it will be in their common intersection, which is perpendicular to the horizontal plane at $P$.

Lay off from the origin of co-ordinates, on the axis of $Z$, a distance $Ap'' = c$, and through $p''$ pass a plane parallel to $YX$: its traces $p''P'$, $p''P''$, will be parallel respectively to the axes $X$ and $Y$, and the point in which the plane is pierced by the perpendicular to the horizontal plane at $P$, will be the position of the required point. The point will therefore be vertically projected on the plane $ZX$ at $P'$, and on the plane $ZY$, at $P''$. Its co-ordinates, are $Pp'$, $pP$, and $pP'$.

The distances of a point from the co-ordinate planes are expressed algebraically by

$$x = a, \qquad y = b, \qquad z = c,$$

and since these conditions determine the position of the point, they are called, the *equations of the point*.

8. Let us now consider these conditions in a general manner, and see what each, taken separately, implies.

The conditions

$$x = \pm a,$$

will limit the point to one of two planes drawn parallel to the co-ordinate plane $YZ$, on different sides of the origin, and at a distance from it equal to $a$.

The conditions

$$y = \pm b,$$

will limit the point to one of two planes drawn parallel to the co-ordinate plane $ZX$, on different sides of the origin, and at a distance from it equal to $b$.

If these conditions exist together, the point will be limited to four straight lines, parallel to the axis of $Z$.

The conditions

$$z = \pm c,$$

will limit the position of the point to one of two planes drawn parallel to the co-ordinate plane $YX$, on different sides of the origin, and at a distance from it equal to $c$.

If all the conditions exist at the same time, the point will be found at either one of the eight points in which the two last planes are pierced by the four parallels before determined; and each of these eight points will be found in one of the eight angles formed by the co-ordinate planes. By attributing to the co-ordinates of these points the signs plus and minus, the position of either one of them may be precisely determined. The following signs are attributed to the co-ordinates of a point in the different angles:

| | | | | |
|---|---|---|---|---|
| 1st | angle | $x = +a,$ | $y = +b,$ | $z = +c,$ |
| 2d | " | $x = -a,$ | $y = +b,$ | $z = +c,$ |
| 3d | " | $x = -a,$ | $y = -b,$ | $z = +c,$ |
| 4th | " | $x = +a,$ | $y = -b,$ | $z = +c,$ |

| | | | | |
|---|---|---|---|---|
| 5th | angle | $x = +a$, | $y = +b$, | $z = -c$, |
| 6th | " | $x = -a$, | $y = +b$, | $z = -c$, |
| 7th | " | $x = -a$, | $y = -b$, | $z = -c$, |
| 8th | " | $x = +a$, | $y = -b$, | $z = -c$. |

9. Since the co-ordinate of a point represents its distance from one of the co-ordinate planes, it follows, that when this distance is 0, the point will be found in the plane.

Hence, we have the following for the equations of the co-ordinate planes:

For the co-ordinate plane $YAX$,

$$z = 0, \quad x \text{ and } y \text{ indeterminate};$$

that is, $x$ and $y$ must be indeterminate in order that they may be made to represent, in succession, the co-ordinates of every point of the plane.

For the co-ordinate plane $XAZ$,

$$y = 0, \quad x \text{ and } z \text{ indeterminate}.$$

For the co-ordinate plane $YAZ$,

$$x = 0 \quad y \text{ and } z \text{ indeterminate}.$$

10. Since either axis lies in two of the co-ordinate planes, we shall have, for the equation of the axis of $X$,

$$y = 0, \quad z = 0, \quad \text{and } x \text{ indeterminate}.$$

For the equation of the axis of $Y$,

$$x = 0, \quad z = 0, \quad \text{and } y \text{ indeterminate}.$$

For the equation of the axis of $Z$,

$$x = 0, \quad y = 0, \quad \text{and } z \text{ indeterminate}.$$

And for the origin, which lies in the three planes,

$$x = 0, \quad y = 0, \quad \text{and} \quad z = 0.$$

**11.** We also have, for a point on the axis of $X$,

$$y=0, \qquad z=0, \qquad \text{and} \qquad x=\pm a.$$

For a point on the axis of $Y$,

$$x=0, \qquad z=0, \qquad \text{and} \qquad y=\pm b.$$

For a point on the axis of $Z$,

$$x=0, \qquad y=0, \qquad \text{and} \qquad z=\pm c.$$

PROPOSITION I. PROBLEM.

*To find the distance between two points in space when their co-ordinates are known.*

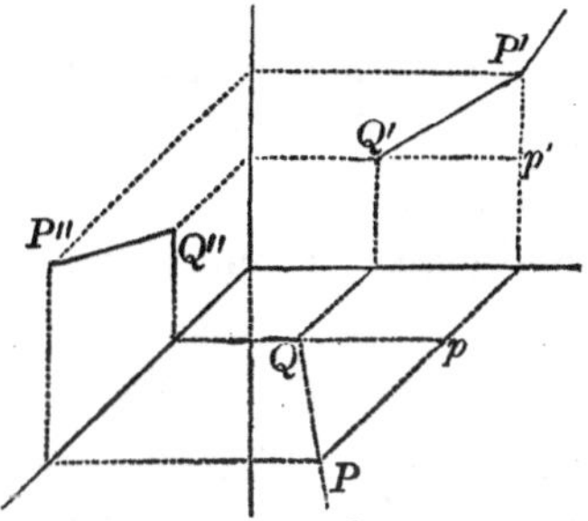


Let $(Q, Q', Q'')$ be one of the points, and $(P, P', P'')$ the other.

Represent the co-ordinates of the first by $x', y', z'$, and those of the second by $x'', y'', z''$, and designate the length of the required line by $D$. The line $D$ will be the hypothenuse of a triangle, of which the base is $QP$, and altitude $p'P'$.

But, $Qp=x''-x'$, $Pp=y''-y'$, and $p'P'=z''-z'$

In the right angle triangle $QPp$, we have

$$\overline{QP}^2=(x''-x')^2+(y''-y')^2,$$

hence, $$D^2=(x''-x')^2+(y''-y')^2+(z''-x')^2$$

and $$D=\sqrt{(x''-x')^2+(y''-y')^2+(z''-z')^2}.$$

*Scholium* 1. If the line were projected on the three co-ordinate axes, $x''-x'$, $y''-y'$, $z''-z'$, would represent, respectively, the length of the projection on each axis: hence, it follows, that *the square of any portion of a straight line is equal to the sum of the squares of its three projections on the co-ordinate axes.*

*Scholium* 2. If one of the points, the one for example of which the co-ordinates are $x', y', z'$, be placed at the origin, we shall have

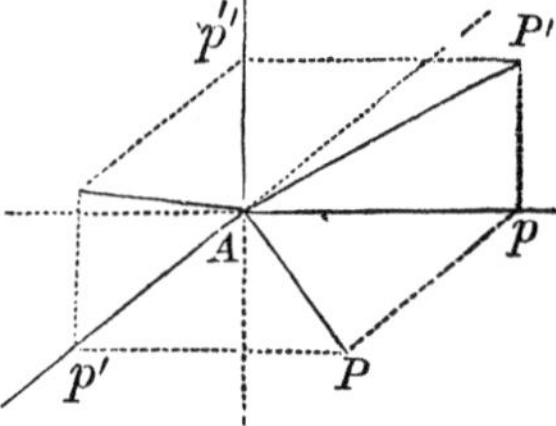


$$D = \sqrt{x''^2 + y''^2 + z''^2},$$

which represents the distance of any point in space from the origin of co-ordinates.

*Scholium* 3. The three lines $Pp'$, $Pp$, $P'p$, drawn perpendicular to the co-ordinate planes, may be regarded as the three edges of a parallelopipedon, of which the line drawn to the origin is the diagonal. We have therefore verified a proposition of geometry, viz: *the sum of the squares of the three edges of a rectangular parallelopipedon is equal to the square of its diagonal.*

*Scholium* 4. This last result offers an easy method of determining a relation that exists between the cosines of the angles which a straight line makes with the co-ordinate axes.

Let us designate the length of the line passing through the origin of co-ordinates by $r$, and the angles which it forms with the axes, respectively, by $X$, $Y$, and $Z$.

We shall then have for the lines $Ap$, $Ap'$, $Ap''$, which are respectively designated by $x''$, $y''$, $z''$, the following values. viz.:

$$x'' = r \cos X, \qquad y'' = r \cos Y, \qquad z'' = r \cos Z.$$

By squaring these equations and adding, we obtain

$$x''^2 + y''^2 + z''^2 = r^2(\cos^2 X + \cos^2 Y + \cos^2 Z).$$

But we have already found

$$x''^2 + y''^2 + z''^2 = r^2.$$

Hence, $$\cos^2 X + \cos^2 Y + \cos^2 Z = 1,$$

that is, *the sum of the squares of the cosines of the three angles which a straight line forms with the three co-ordinate axes, is equal to radius square, or unity.*

### PROPOSITION II. PROBLEM.

*To find the equations of a straight line in space*

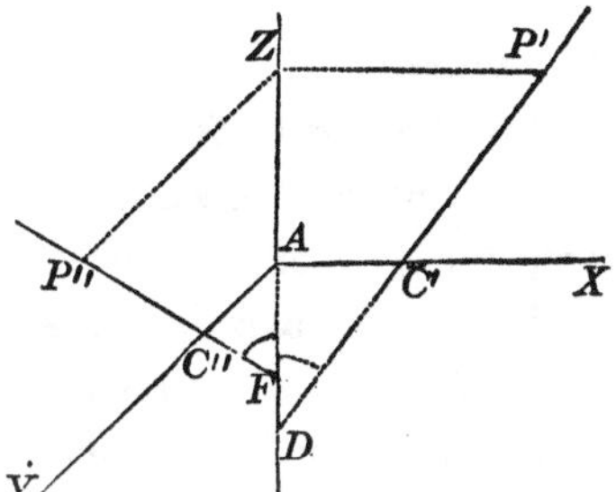


Let $C'P'$ be the projection of a straight line on the co-ordinate plane $ZX$, and $C''P''$ its projection on the co-ordinate plane $YZ$. Now, since a line is determined in space when two of its projections are known (Des. Geom. Art. 26), it follows that the conditions which fix the projections will determine the line.

Let $$x = az + \alpha$$

be the equation of the projection $C'P'$, and

$$y = bz + \beta,$$

the equation of the projection $C''P''$.

In these equations, $a$ represents the tangent of the angle $ADP'$, $\alpha$ the distance $AC'$, $b$ the tangent of the angle $P''FZ$, and $\beta$ the distance $AC''$. The angles in the co-ordinate plane $ZX$, are estimated from the axis $Z$ to the right, and in the co-ordinate plane $YZ$, they are estimated from the axis $Z$ towards the left.

If we suppose $a$, $\alpha$, $b$, and $\beta$, to be given or known, the two projections $C'P'$, $C''P''$ will be determined; and hence, the line of which they are the projections will be determined in space. Hence,

$$x = az + \alpha, \qquad y = bz + \beta,$$

are the equations of a straight line.

*Scholium* 1. Since the projections of a straight line on two of the co-ordinate planes determine the position of the line in space, they ought also to determine its projection on the third co-ordinate plane. This indeed may be easily verified.

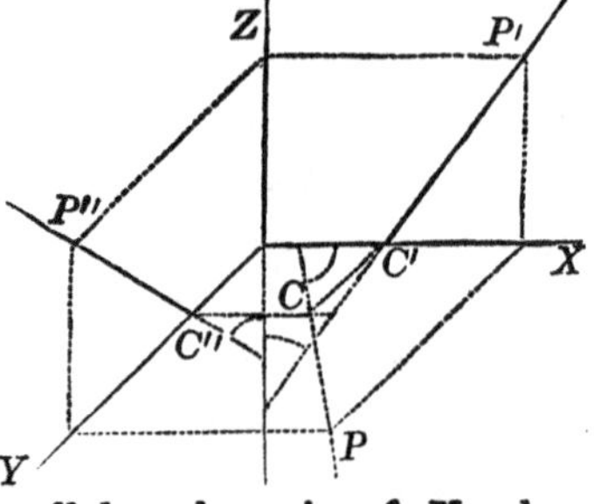


For, through $P'$ draw a parallel to the axis of $Z$, and from the point in which it intersects the axis of $X$ draw a parallel to the axis of $Y$. Through $P''$ draw a parallel to the axis of $Z$, and through the point in which it intersects the axis of $Y$, draw a parallel to the axis of $X$: then will $P$ be the projection of the point ($P'$, $P''$), on the co-ordinate plane $YX$.

Find, in a similar manner, the projection of a second point, as ($C'$, $C''$), and draw the projection $CP$.

The equations

$$x = az + \alpha, \qquad y = bz + \beta,$$

of the projections of the line on two of the co-ordinate planes, ought also to give the equation of the projection on the third plane.

If we eliminate $z$ from the two last equations, we have

$$\frac{x-\alpha}{a}=\frac{y-\beta}{b}, \quad \text{or} \quad \frac{b}{a}=\frac{y-\beta}{x-\alpha},$$

hence, $$y-\beta=\frac{b}{a}(x-\alpha).$$

This equation is independent of $z$, and represents the projection $CP$ of the line, on the third co-ordinate plane $YX$.

*Scholium* 2. If in the equations of a given straight line,

$$x=az+\alpha, \qquad y=bz+\beta,$$

a particular value be attributed to either of the variables $x$, $y$, or $z$, the corresponding values of the two others may be found from the equations, since there will then be but two unknown quantities in the two equations.

The corresponding values of the co-ordinates may likewise be found in the third equation,

$$y-\beta=\frac{b}{a}(x-\alpha).$$

The geometrical construction will also correspond to these results. For, if we assume any point, as $P'$, in one of the projections of the line, the two other projections, $P$ and $P''$, may easily be constructed.

*Scholium* 3. Let us now consider the equations

$$x=az+\alpha, \qquad y=bz+\beta,$$

separately.

The equation

$$x=az+\alpha,$$

being independent of $y$, will be satisfied for every point of the plane passing through $C'P'$, and perpendicular to the co-ordinate plane $ZX$: hence, it may be regarded as the equation of that plane.

In like manner, the equation

$$y = bz + \beta,$$

being independent of $x$, will be satisfied for every point in the plane passing through $C''P''$, and perpendicular to the co-ordinate plane $YZ$: hence, it may be regarded as the equation of that plane.

For similar reasons

$$y - \beta = \frac{b}{a}(x - \alpha),$$

is the equation of a plane passing through $CP$, and perpendicular to the co-ordinate plane $YX$.

*Scholium* 4. Let us now consider the conditions which would be imposed upon the straight line, by supposing $a$, $b$, $\alpha$, and $\beta$, to become known in succession.

When $a$, $b$, $\alpha$, and $\beta$, are all undetermined, the equations

$$x = az + \alpha, \qquad y = bz + \beta,$$

may be made to represent every straight line which can be drawn in space, by attributing suitable values to $a$, $b$, $\alpha$, and $\beta$. And when $a$, $b$, $\alpha$, and $\beta$, have given values, the equations will designate but a single straight line.

If we suppose $a$ to be given, the line may have any position in space, such that its projection on the co-ordinate plane $ZX$ shall make an angle with $Z$, of which the tangent is $a$.

If we suppose $\alpha$ also to be given, the projection of the line on the co-ordinate plane $ZX$, will intersect the axis of $X$ at a given point. and the two conditions will limit the line to a given plane. Its position in the plane will still be entirely undetermined.

If we now suppose $b$ to be given, the direction of the line will then be determined, but it may still have an indefinite number of parallel positions in the given plane.

If, finally, we attribute a value to $\beta$, the projection on the plane of $YZ$ will intersect the axis of $Y$ at a given point; and hence, the position of the line will become known. The letters $\alpha$ and $\beta$ represent the co-ordinates of the points in which the line intersects the co-ordinate plane $YX$.

The resolution of problems involving the straight line in space, consists in finding such values for the arbitrary constants $a$, $b$, $\alpha$, and $\beta$, as shall satisfy the required conditions.

*Scholium* 5. Before, however, proceeding to the resolution of problems, it may be well to remark, that the same methods which have been adopted for determining straight lines in space, by means of the equations of their projections on two of the co-ordinate planes, are equally applicable to curves.

If, for example, a curve be projected on the co-ordinate plane $ZX$, the equation of the projection will contain the variables $x$ and $z$, and will be independent of $y$. The equation of the projection will therefore be satisfied for any point of the surface of the right cylinder which projects the curve.

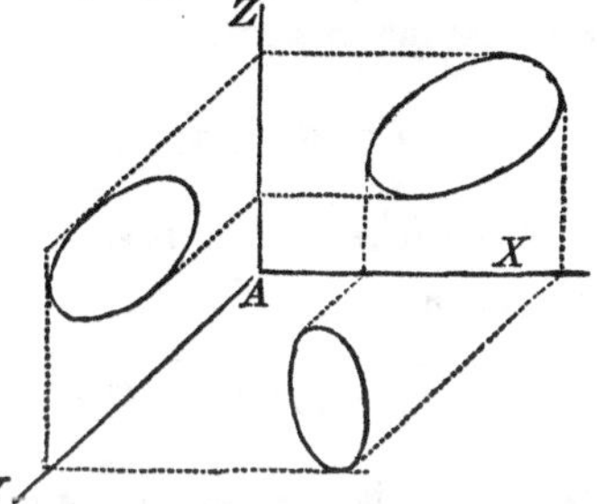


If the curve be now projected on the co-ordinate plane $YZ$, the equation of the projection will be independent of $x$, and will therefore be satisfied for every point of the surface of the right cylinder which projects the curve on $YZ$.

The two equations together will determine two right cylinders whose bases are the curves on the co-ordinate planes, and which, by their intersection, will determine the curve in space.

If $z$ be eliminated from the two equations, we shall have the equation of the projection of the curve on the third co-ordinate plane.

PROPOSITION III. PROBLEM.

*To find the equations of a straight line which shall pass through two given points.*

Let $x'$, $y'$, $z'$, and $x''$, $y''$, $z''$, be the co-ordinates of the given points.

The required equations will be of the form

$$x = az + \alpha, \quad (1) \qquad y = bz + \beta, \quad (2)$$

in which it is required to find such values for $a$, $\alpha$, $b$, and $\beta$, as shall cause the right line to fulfil the required conditions.

Since the straight line is to pass through a point of which the co-ordinates are $x'$, $y'$, $z'$, we shall have

$$x' = az' + \alpha, \quad (3) \qquad y' = bz' + \beta; \quad (4)$$

and since it is also to pass through a point of which the co ordinates are $x''$, $y''$, $z''$, we shall likewise have

$$x'' = az'' + \alpha, \quad (5) \qquad y'' = bz'' + \beta. \quad (6)$$

The four last equations enable us to determine the four constants $a$, $\alpha$, $b$, $\beta$.

By subtracting the fifth equation from the third, and the sixth from the fourth, we obtain

$$x' - x'' = a(z' - z'') \quad \text{and} \quad y' - y'' = b(z' - z''),$$

from which we find,

$$a = \frac{x' - x''}{z' - z''}, \quad \text{and} \quad b = \frac{y' - y''}{z' - z''},$$

hence, $a$ and $b$ are determined, and if their known values be substituted respectively in equations (3) and (4), or (5) and (6), the values of $\alpha$ and $\beta$, will become known.

If we subtract the third equation from the first, and the fourth from the second, we shall have

$$x - x' = a(z - z') \qquad\qquad y - y' = b(z - z'),$$

which are the equations of a straight line passing through a given point. Substituting for $a$ and $b$ their known values, and we have

$$x - x' = \frac{x' - x''}{z' - z''}(z - z') \qquad y - y' = \frac{y' - y''}{z' - z''}(z - z'),$$

which are the equations of a straight line passing through the two given points.

### PROPOSITION IV. PROBLEM.

*To find the conditions which will cause a straight line to be parallel to a given straight line.*

Let

$$x = az + \alpha, \qquad\qquad y = bz + \beta,$$

be the equations of the given line; and

$$x = a'z + \alpha', \qquad\qquad y = b'z + \beta',$$

the equations of the required line.

The two lines will be parallel in space when their projections on two of the co-ordinate planes are parallel (Des. Geom. Art 30). The projections will be rendered parallel by making

$$a' = a \qquad \text{and} \qquad b' = b,$$

hence, the equations of the required line will become

$$x = az + \alpha', \qquad\qquad y = bz + \beta';$$

and since $\alpha'$ and $\beta'$ are yet undetermined, there is an infinite number of lines which will fulfil the conditions.

If we make

$$\alpha' = \alpha \quad \text{and} \quad \beta' = \beta,$$

the two lines will coincide.

### PROPOSITION V. PROBLEM.

*To find the angle included between two lines given by their equations, in terms of the angles which the lines make with the co-ordinate axes.*

Let $$x = az + \alpha, \qquad y = bz + \beta,$$

be the equations of the first line, and

$$x = a'z + \alpha', \qquad y = b'z + \beta',$$

be the equations of the second.

It has been observed (Geom. Bk. VI, Prop. VI, Sch.), that two straight lines which cross each other in space, may be regarded as forming an angle, although they do not lie in the same plane. They are supposed to make the same angle with each other as would be formed by one of the lines, and a line drawn through any point of it and parallel to the other: or as would be formed by two lines drawn through the same point and respectively parallel to the given lines.

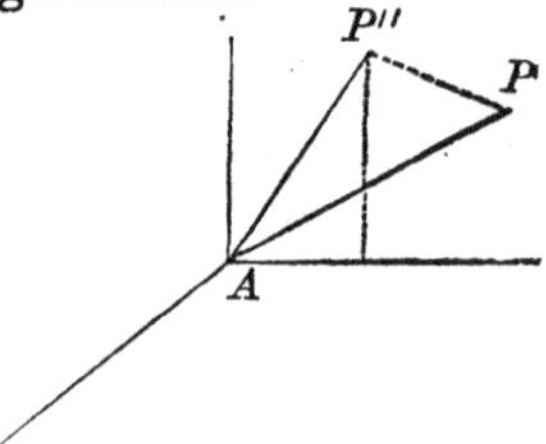


If then, two lines be drawn through the origin of co-ordinates respectively parallel to the given lines, the angle which they form with each other will be equal to the required angle.

The equation of these lines will be

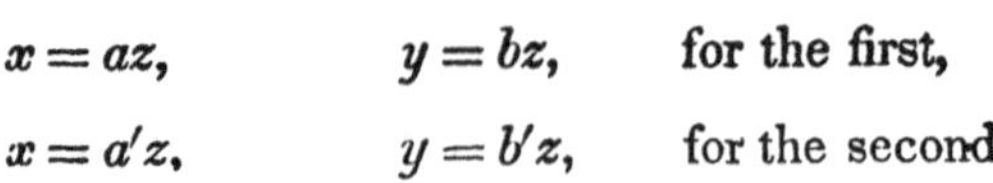

$$x = az, \qquad y = bz, \qquad \text{for the first,}$$

$$x = a'z, \qquad y = b'z, \qquad \text{for the second}$$

Let us take on the first line any point as $P'$ and designate its co-ordinates by $x'$, $y'$, $z'$, and its distance from the origin by $r'$. Take, in like manner, on the second line any point as $P''$, and designate its co-ordinates by $x''$, $y''$, $z''$, and its distance from the origin by $r''$, and let $D$ denote the distance between the points. If we designate the angle included between the lines by $V$, we shall have, in the triangle $AP'P''$ (Trig. Th. IV).

$$\cos V = \frac{r'^2 + r''^2 - D^2}{2r'r''},$$

and we have now only to find $r'$, $r''$ and $D$.

Let us designate the three angles which the first line forms with the co-ordinate axes respectively, by $X$, $Y$ and $Z$, and the angles which the second line forms with the same axes by $X'$, $Y'$ and $Z'$; we shall then have (Prop. I, Sch. 4),

$$x' = r' \cos X, \qquad y' = r' \cos Y, \qquad z' = r' \cos Z,$$

$$x'' = r'' \cos X', \qquad y'' = r'' \cos Y', \qquad z'' = r'' \cos Z'.$$

But the square of the distance between two points is (Prop. I),

$$D^2 = (x' - x'')^2 + (y' - y'')^2 + (z' - z'')^2, \qquad \text{or,}$$

$$D^2 = x'^2 + y'^2 + z'^2 + x''^2 + y''^2 + z''^2 - 2(x'x'' + y'y'' + z'z'');$$

or by substituting for the co-ordinates of the points their distances from the origin into the cosines of the angles which the lines make with the co-ordinate axes, we have,

$$D^2 = \left\{ \begin{array}{c} r'^2 (\cos^2 X + \cos^2 Y + \cos^2 Z) + r''^2 (\cos^2 X' + \cos^2 Y' + \cos^2 Z') \\ - 2r'r'' (\cos X \cos X' + \cos Y \cos Y' + \cos Z \cos Z') \end{array} \right\}$$

But it has been shown (Prop. I, Sch. 4), that,

$$\cos^2 X + \cos^2 Y + \cos^2 Z = 1, \quad \cos^2 X' + \cos^2 Y' + \cos^2 Z' = 1$$

and hence,

$$D^2 = r'^2 + r''^2 - 2r'r''(\cos X \cos X' + \cos Y \cos Y' + \cos Z \cos Z').$$

If this value of $D^2$ be substituted in the equation,

$$\cos V = \frac{r'^2 + r''^2 - D^2}{2r'r''},$$

we shall find, after dividing by $2r'r''$,

$$\cos V = \cos X \cos X' + \cos Y \cos Y' + \cos Z \cos Z',$$

that is, *the cosine of the angle included between two lines is equal to the sum of the rectangles of the cosines of the angles which the lines in space form with the co-ordinate axes.*

*Second Method.*

Having found

$$\cos V = \frac{r'^2 + r''^2 - D^2}{2r'r''},$$

as in the first method, we may place it under the form

$$D^2 - r'^2 - r''^2 + 2r'r'' \cos V = 0.$$

We next find the value of $D^2$, as before, viz:

$$D^2 = \left\{ \begin{array}{l} r'^2(\cos^2 X + \cos^2 Y + \cos^2 Z) + r''^2(\cos^2 X' + \cos^2 Y' + \cos^2 Z') \\ \quad -2r'r''(\cos X \cos X' + \cos Y \cos Y' + \cos Z \cos Z') \end{array} \right\}.$$

Substituting this value of $D^2$ in the last equation, it may be placed under the form

$$\left. \begin{array}{l} r'^2(\cos^2 X + \cos^2 Y + \cos^2 Z - 1) + r''^2(\cos^2 X' + \cos^2 Y' + \cos^2 Z' - 1) \\ \quad -2r'r''(\cos X \cos X' + \cos Y \cos Y' + \cos Z \cos Z' - \cos V) \end{array} \right\} = 0.$$

Now, since the angle $V$ which the two lines make with each other, and the angles which they form with the co-ordinate axes, are entirely independent of the distances $r'$ and $r''$, it follows, that the coefficients in the last equation will be independent of the values which may be attributed to $r'$ and $r''$.

But this equation will be true whatever values may be attributed to $r'$ and $r''$: hence, the coefficients of the like powers of $r'$ and $r''$, must be separately equal to 0 (Alg Art. 208).

This condition will give

$$\cos^2 X + \cos^2 Y + \cos^2 Z = 1, \quad \cos^2 X' + \cos^2 Y' + \cos^2 Z' = 1,$$

and $\cos V = \cos X \cos X' + \cos Y \cos Y' + \cos Z \cos Z'$.

The first two equations prove the same property as was proved in (Prop. I, Sch. 4), and employed in the first part of this proposition. The third equation gives the same value for $\cos V$ as before found.

The second method of determining the cosine of the angle included between the lines, affords a striking and elegant application of the method of indeterminate coefficients.

*Scholium* 1. Having found the cosine of the angle included between two lines in terms of the angles which they form with the co-ordinate axes in space, we shall, in the next place, find the same value in terms of the angles which the projections of the lines on the co-ordinate planes $ZX$, $YZ$, form with the axis of $Z$.

The equations of the parallels through the origin, are

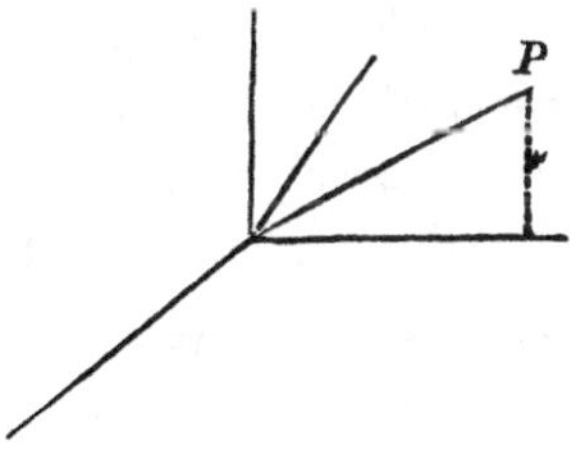

$$x = az, \qquad y = bz,$$
$$x = a'z, \qquad y = b'z,$$

Let us designate the co-ordinates of the point $P$, on the first line, by $x'$, $y'$, $z'$, we shall then have

$$x' = az', \qquad y' = bz',$$

and for the value of $r'$,

$$r'^2 = x'^2 + y'^2 + z'^2.$$

From these three equations we find,

$$x' = \frac{ar'}{\sqrt{1+a^2+b^2}}, \quad y' = \frac{br'}{\sqrt{1+a^2+b^2}}, \quad z' = \frac{r'}{\sqrt{1+a^2+b^2}}$$

But we have already found (Prop. I, Sch. 4),

$$x' = r' \cos X, \qquad y' = r' \cos Y, \qquad z' = r' \cos Z,$$

Substituting these values and dividing, we obtain

$$\cos X = \frac{a}{\sqrt{1+a^2+b^2}}, \cos Y = \frac{b}{\sqrt{1+a^2+b^2}}, \cos Z = \frac{1}{\sqrt{1+a^2+b^2}}$$

If we reason in the same manner on the equations of the second straight line, we shall find

$$\cos X' = \frac{a'}{\sqrt{1+a'^2+b'^2}}, \quad \cos Y' = \frac{b'}{\sqrt{1+a'^2+b'^2}}, \quad \cos Z' = \frac{1}{\sqrt{1+a'^2+b'^2}}$$

If these values be now substituted in the equation which gives the value of cos $V$, it will reduce to,

$$\cos V = \frac{1 + aa' + bb'}{\pm\sqrt{1+a^2+b^2}\sqrt{1+a'^2+b'^2}}.$$

The cos $V$ will be plus or minus, according as we take the signs of the radicals in the denominator, like or unlike.

The plus value of cos $V$ will correspond to the acute angle, and the minus value to the obtuse angle.

*Scholium* 2. It is evident that the angle which a straight line forms with the axis of either of the co-ordinate planes, is the complement of the angle which the line forms with the plane itself. Hence, if we designate the angles which the first line forms with the co-ordinate planes, respectively, by $u$, $u'$, $u''$, we shall have

$$\sin u = \frac{a}{\sqrt{1+a^2+b^2}}, \sin u' = \frac{b}{\sqrt{1+a^2+b^2}}, \sin u'' = \frac{1}{\sqrt{1+a^2+b^2}}$$

*Scholium* 3. If in the equation

$$\cos V = \frac{1 + aa' + bb'}{\sqrt{1 + a^2 + b^2}\sqrt{1 + a'^2 + b'^2}},$$

we make $V = 90°$, the cos $V$ will become 0; and hence,

$$1 + aa' + bb' = 0,$$

which is the equation of condition by which two right lines are rendered perpendicular to each other in space.

*Scholium* 4. If we make $V = 0$, the two straight lines will become parallel; and the equation will become

$$\pm 1 = \frac{1 + aa' + bb'}{\sqrt{1 + a^2 + b^2}\sqrt{1 + a'^2 + b'^2}}.$$

Squaring both members, clearing the equation of fractions, and reducing, we obtain

$$(a' - a)^2 + (b' - b)^2 + (ab' - a'b)^2 = 0.$$

Each term, in the first member of this equation, being a square, will be positive; and hence, the equation can only be satisfied when the terms are separately equal to 0. These conditions give

$$a' = a, \qquad b' = b, \qquad \text{and} \qquad ab' = a'b;$$

but the third is a dependant condition, being a consequence of the first two.

The first two conditions require, that the projections of the lines on each of the co-ordinate planes $ZX$, $YZ$, be parallel to each other: and hence, they cause the lines to be parallel in space. These conditions are the same as determined in (Prop. IV).

*Scholium* 5. If we suppose either of the lines, the second for example, to coincide in succession with each of the co-ordinate axes, the angle $V$ will, under each supposition become the angle which the first line forms with one of the axes.

Let us suppose, in the first place, that the second line, of which the equations are

$$x = a'z, \qquad y = b'z,$$

is made to coincide with the axis of $Z$, of which the equations are

$$x = 0, \qquad y = 0, \qquad z \text{ indeterminate.}$$

Introducing these values into the equations of the line, we have

$$0 = a'z, \qquad 0 = b'z;$$

and since $z$ is indeterminate,

$$a' = 0, \qquad \text{and} \qquad b' = 0.$$

If these values be substituted in the value of cos $V$, we shall have,

$$\cos V = \frac{1}{\sqrt{1 + a^2 + b^2}} = \cos Z$$

Let us now suppose the second line, whose equations can be put under the forms

$$z = \frac{1}{a'}x, \qquad y = \frac{b'}{a'}x,$$

to coincide with the axis of $X$, of which the equations are

$$z = 0, \qquad y = 0, \qquad x \text{ indeterminate.}$$

Introducing these values into the equation of the line, we find

$$\frac{1}{a'} = 0, \qquad \text{and} \qquad \frac{b'}{a'} = 0$$

The general value for the angle may be placed under the form

$$\cos V = \frac{\frac{1}{a'} + a + \frac{bb'}{a'}}{\sqrt{1 + a^2 + b^2}\sqrt{\frac{1}{a'^2} + 1 + \frac{b'^2}{a'^2}}};$$

hence,

$$\cos V = \frac{a}{\sqrt{1 + a^2 + b^2}} = \cos X.$$

In a similar manner we may find

$$\cos V = \frac{b}{\sqrt{1 + a^2 + b^2}} = \cos Y.$$

These values agree with those before found, and the sum of their square is equal to unity.

## PROPOSITION VI. PROBLEM.

*To find the conditions which will cause two straight lines to intersect in space, and to find the co-ordinates of their point of intersection.*

Let $\qquad x = az + \alpha, \qquad y = bz + \beta,$

and $\qquad x = a'z + \alpha', \qquad y = b'z + \beta',$

be the equations of the lines, in which we will at first suppose the arbitrary constants to be undetermined.

If these lines intersect each other in space, they must have one point in common, and the co-ordinates of this point will satisfy the equations of both the lines. If we designate the co-ordinates of the common point by $x'$, $y'$, $z'$, we shall have

$$x' = az' + \alpha, \qquad y' = bz' + \beta,$$

$$x' = a'z' + \alpha' \qquad y' = b'z' + \beta'.$$

Eliminating $x'$ and $y'$ from these equations, we find

$$(a-a')z'+\alpha-\alpha'=0, \qquad (b-b')z'+\beta-\beta'=0:$$

and if $z'$ be eliminated from the two last equations, we have

$$(a-a')(\beta-\beta')-(\alpha-\alpha')(b-b')=0,$$

which is called the *equation of condition*, since it must always be satisfied in order that the two straight lines may intersect each other.

There are eight arbitrary constants entering into this equation. It may therefore be satisfied in an infinite number of ways. Indeed, if values be attributed at pleasure to seven of the constants, such a value may, in general, be found for the remaining one as will satisfy the equation, and consequently cause the lines to intersect each other.

Let us now consider the second part of the proposition, viz.: to find the co-ordinates of the point of intersection when the lines meet each other. We find from the previous equations

$$z'=\frac{\alpha'-\alpha}{a-a'}, \qquad \text{or} \qquad z'=\frac{\beta'-\beta}{b-b'}.$$

and

$$x'=\frac{a\alpha'-a'\alpha}{a-a'}, \qquad y'=\frac{b\beta'-b'\beta}{b-b'}.$$

These values of the co-ordinates of the point of intersection become infinite when

$$a=a' \qquad \text{and} \qquad b=b';$$

but these conditions cause the lines to become parallel to each other, and hence, their point of intersection ought to be at an infinite distance from the origin of co-ordinates.

If we have, at the same time,

$$\alpha'=\alpha \qquad \text{and} \qquad \beta'=\beta,$$

the co-ordinate of the point of intersection will become $\frac{0}{0}$, or indeterminate; as indeed they should do, since the two lines would then coincide throughout their whole extent

*Scholium* 1. The condition of intersection

$$(a-a')(\beta-\beta')-(\alpha-\alpha')(b-b')=0,$$

is independent of the condition

$$1+aa'+bb'=0,$$

which causes two straight lines to be perpendicular to each other in space. The second condition may, therefore, be satisfied independently of the first; and hence, two straight lines may be perpendicular in space without intersecting.

*Scholium* 2. The same principles may be readily employed in determining the intersection of two curves given by their projections.

Let
$$4z^2-8z+x^2-2x=-4, \quad (1)$$
$$4z^2-8z+y^2-4y=-7, \quad (2)$$

be the equations of one of the curves, and

$$z^2-2z-2x=-3, \quad (3)$$
$$z^2-2z-2y=-5, \quad (4)$$

the equations of the other.

If we combine the first and third equations by eliminating $z$, the resulting equation will determine the abscissas of all the points common to the two curves; and it may also give values of $x$ which do not correspond to points of intersection. To determine which of the roots correspond to points of intersection, let each be substituted, in succession, in equations (1) and (3), and those which give equal values of $z$ in the two equations, will correspond to the points in which the projections of the curves intersect each other, and the ordinate $z$ of the points of intersection will thus become known

18

Combine, in a similar manner, equations (2) and (4), and find the values of the ordinates $z$, in which the projections of the curves on the plane of $YZ$ intersect each other.

Then compare the values of $z$ found for the projections on the co-ordinate plane $ZX$ with those found for the co-ordinate plane $YZ$, and all the equal values will indicate the points at which the curves intersect in space. For, in order that two curves may intersect in space, it is not only necessary that their projections intersect each other, but the corresponding points of intersection must lie in the same perpendicular to the common intersection of the two planes on which their projections are made: and hence, the ordinates of these points, estimated in parallels to this axis, must be the same for the four curves into which the given curves are projected.

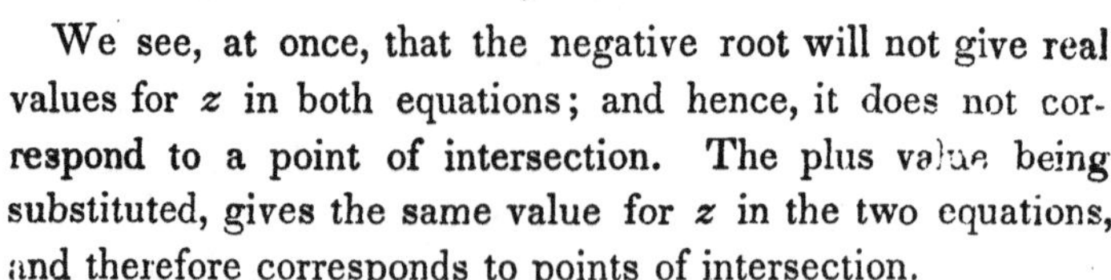

In eliminating $z$ from equations (1) and (3), we find in the resulting equation

$$x^2 + 6x = 8,$$

which gives

$$x = -3 + \sqrt{17},$$

and $$x = -3 - \sqrt{17}.$$

We see, at once, that the negative root will not give real values for $z$ in both equations; and hence, it does not correspond to a point of intersection. The plus value being substituted, gives the same value for $z$ in the two equations, and therefore corresponds to points of intersection.

If we now combine equations (2) and (4), and eliminate $z$, we shall find

$$y = -2 + \sqrt{17}, \quad \text{and} \quad y = -2 - \sqrt{17},$$

in which the negative root is again to be rejected, and the positive root gives the same values for $z$ in equations (2) and (4) as already found in equations (1) and (3). Hence, the curves intersect each other in two points, of which the co-ordinates are

$$x=-3+\sqrt{17}, \quad y=-2+\sqrt{17}, \quad z=1+\sqrt{-8+2\sqrt{17}},$$

and

$$x=-3+\sqrt{17}, \quad y=-2+\sqrt{17}, \quad z=1-\sqrt{-8+2\sqrt{17}}.$$

*Scholium* 3. There is yet another method of determining the points of intersection of two curves given by their projections, and which, in a purely analytical sense is more elegant than the one just considered. It will, however, in its practical applications, generally be found more difficult. The method is this:

If we eliminate $x$ from equations (1) and (3), the resulting equation will be expressed in terms of $z$ and constant quantities. If $y$ be eliminated from equations (2) and (4), the resulting equation will, in like manner, be expressed in terms of $z$ and constants.

If we now find the greatest common divisor of these two equations and place it equal to 0, the roots of the equation thus obtained will give all the values of $z$, which will satisfy at the same time the four equations that represent the projections of the given curves; and it may, also, give other roots. Substitute these roots, in succession, in each of the four equations and any root that will give for $x$ real and equal values in equations (1) and (3), and for $y$ real and equal values in equations (2) and (4), will correspond to a point of intersection of the curves in space.

The last method has this advantage over the other. It explains the manner of causing two curves to intersect each other when values can be assigned at pleasure to the arbitrary constants which determine their position in space. For, if the constants are undetermined such values may be attributed to them as shall not only give a common divisor in $z$, but as will, also, cause the roots of that common divisor, when it is placed equal to 0, to fulfil the other conditions of intersection

## *Of the Plane.*

The equation of a plane is an equation expressing the relations between the co-ordinates of every point of the plane.

### PROPOSITION VII. PROBLEM.

*To find the equation of a plane.*

A line is said to be perpendicular to a plane when it is perpendicular to every line passing through its foot and lying in the plane: and conversely the plane is said to be perpendicular to the line (Geom. Bk. VI, Def. 1).

A plane may, therefore, be generated or described by drawing a line perpendicular to a given line, and then permitting the perpendicular to revolve about the point of intersection. If the perpendicular be at right angles to the given line, in all its positions, it will describe a plane surface.

Let $$x = az + \alpha, \qquad y = bz + \beta,$$

be the equations of a given line.

If we designate the co-ordinates of a particular point by $x'$, $y'$, $z'$, the equations of the line passing through this point, will be,

$$x - x' = a(z - z'), \qquad y - y' = b(z - z').$$

The equations of a perpendicular passing through the given point, of which the co-ordinates are $x'$, $y'$, $z'$, are

$$x - x' = a'(z - z'),$$
$$y - y' = b'(z - z').$$

But the equation of condition which causes two lines to be at right angles is

$$1 + aa' + bb' = 0.$$

If we now attribute to $a'$ and $b'$ all possible values which will satisfy this equation, we shall have, in succession, all the perpendiculars which can be drawn to the given line through the given point of which the co-ordinates are $x'$, $y'$, $z'$, and these perpendiculars determine the plane.

It is necessary, however, to find the equation of the plane in terms of the co-ordinates of its different points. We find from the equations of the perpendicular

$$a' = \frac{x - x'}{z - z'}, \qquad b' = \frac{y - y'}{z - z'}.$$

Substituting these values in the equation of condition,

$$1 + aa' + bb' = 0,$$

and reducing, we find

$$z - z' + a(x - x') + b(y - y') = 0:$$

but, since $a$, $b$, $z'$, $x'$, $y'$, are known quantities, we may represent the constant part of the equation by a single letter, by making

$$-z' - ax' - by' = -c;$$

hence, the equation of the plane becomes

$$z + ax + by - c = 0.$$

*Scholium* 1. Since the equation of the plane contains

three variables we may assign values at pleasure to two of them, and the equation will then make known the value of the third. For example, if we assign known values $x'$ and $y$ to $x$ and $y$, the equation of the plane will give

$$z = c - ax' - by',$$

and hence the co-ordinate $z$ becomes known.

*Scholium* 2. The lines in which a plane intersects the co-ordinate planes, are called the *traces of the plane.* These traces are found by combining the equation of the plane with the equations of the co-ordinate planes.

Thus, if in the equation

$$z + ax + by - c = 0,$$

we make $y = 0$, which is the characteristic of the co-ordinate plane $ZX$, the resulting equation,

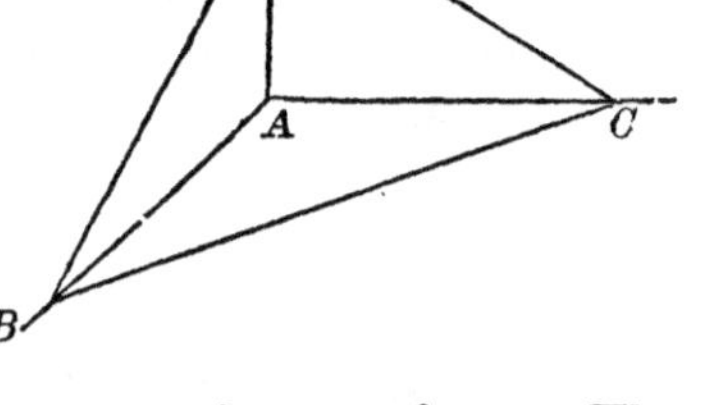


$$z + ax - c = 0,$$

will designate the trace $CD$ common to the two planes. The equation may be placed under the form

$$z = -ax + c;$$

and hence, the trace may be drawn. Or if we make, in succession,

$$x = 0, \quad \text{and} \quad z = 0,$$

we shall find

$$z = c = AD, \quad \text{and} \quad x = \frac{c}{a} = AC,$$

and the trace may then be drawn through the points $C$ and $D$

We likewise find, for the trace $BD$,

$$z = -by + c:$$

and for the trace $BC$, $y = -\frac{a}{b}x + \frac{c}{b}$.

We also find $AD = c$ by making $y = 0$ in the equation of the trace $BD$, and $AB = \frac{c}{b}$ by making $x = 0$ in the equation of the trace $BC$, or by making $z = 0$ in the equation of the trace $BD$.

By comparing the equations of the traces with the equation in Bk. II, Prop. II, we see that,

$-a$ is the tangent of the angle which the trace $CD$ makes with the axis of $X$:

$-b$, the tangent of the angle which the trace $BD$ makes with the axis of $Y$: and

$-\frac{a}{b}$, the tangent of the angle which the trace $BC$ makes with the axis of $X$.

*Scholium* 3. The equations of the straight line, to which the plane has been drawn perpendicular, are

$$x - x' = a(z - z'), \qquad y - y' = b(z - z').$$

and the equations of the traces $CD$, $BD$ may be placed under the form

$$x = -\frac{1}{a}z + \frac{c}{a}, \quad y = -\frac{1}{b}z + \frac{c}{b}.$$

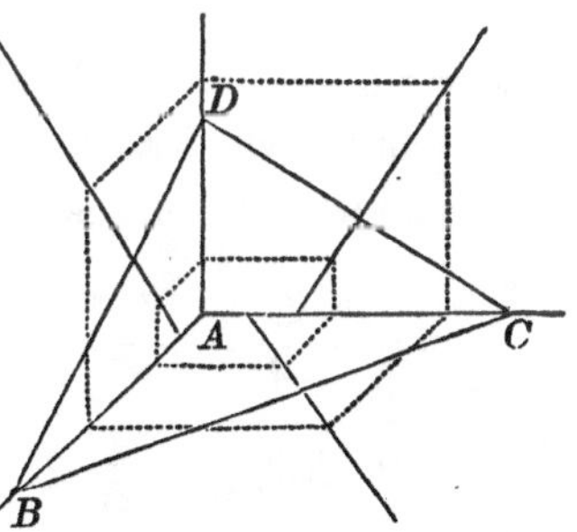


By comparing the coefficient of $z$, in the equation of the projection of the line on the co-ordinate plane $ZX$, with the coefficient of $z$ in the equation of the trace $CD$, we find that their product plus unity is equal to 0: hence, the lines are at right angles to each other. The same may be shown for the trace $BD$, and the projection on the plane $YZ$; and also for the trace $BC$, and the projection on the plane $YX$.

This verifies a well known property, viz.: *If a line be perpendicular to a plane in space, the projections of the line will be respectively perpendicular to its traces.*

*Scholium* 4. The equation of a plane may be written under the form

$$Ax + By + Cz + D = 0,$$

in which $A$, $B$, $C$, and $D$, are constant for the same plane, but have different values when the equation represents different planes. The coefficients $A$, $B$, and $C$, are functions of the angles which the traces of the plane form with the co-ordinate axes, and $D$ is a function of the distances from the origin to the points in which the plane cuts the co-ordinate axes. If the plane passes through the origin of co-ordi nates, its equation takes the form

$$Ax + By + Cz = 0.$$

The plane whose equation is

$$Ax + By + Cz + D = 0,$$

may be readily constructed by finding its traces on the co-ordinate planes, which is done by making the variables $x$, $y$, and $z$ equal to 0 in succession: we have for the trace $CD$,

$$x = -\frac{C}{A}z - \frac{D}{A};$$

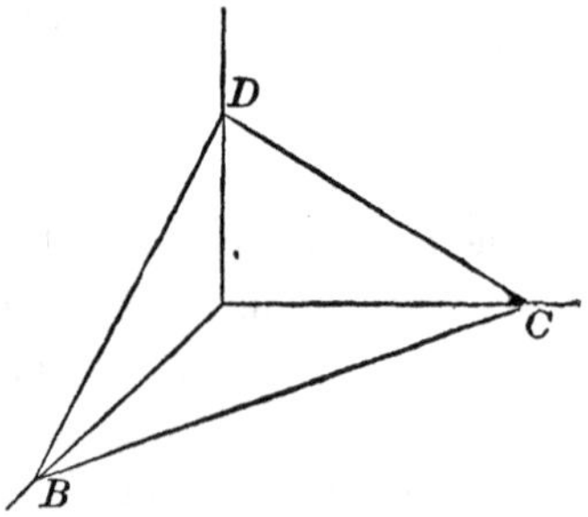

for the trace $BD$,

$$y = -\frac{C}{B}z - \frac{D}{B};$$

for the trace $BC$

$$y = -\frac{A}{B}x - \frac{D}{B}.$$

*Scholium* 5. If the plane is perpendicular to the plane $YX$, its trace $CD$, will be parallel to the axis of $Z$: and hence, the tangent of the angle which it makes with $Z$ will be 0: this will give

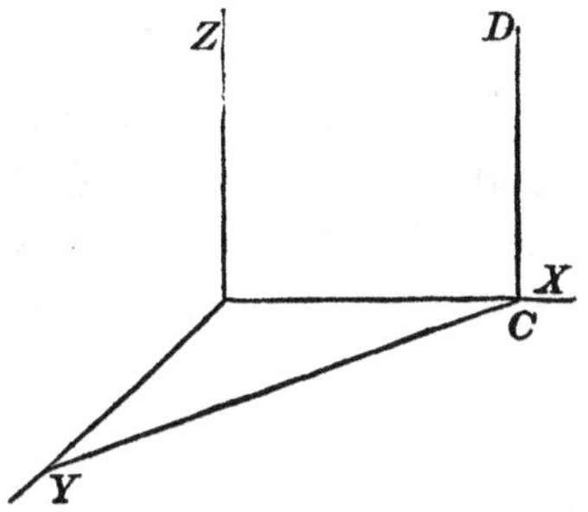

$$\frac{C}{A}=0,$$

and consequently $C=0$.

Substituting this value in the equation of the plane, and it becomes

$$Ax+By+D=0.$$

If the plane be perpendicular to $ZX$, we shall find, by similar reasoning, $B=0$, and the equation of the plane will reduce to

$$Ax+Cz+D=0.$$

If we suppose the plane to be perpendicular to the plane $YZ$, we shall find $A=0$, and the equation of the plane will become

$$By+Cz+D=0.$$

## PROPOSITION VIII. THEOREM.

*Every equation of the first degree, between three variables, is the equation of a plane.*

A plane is a surface, in which, if two points be assumed at pleasure, and connected by a straight line, that line will lie wholly in the surface (Geom. Bk. I, Def. 6). If then, it be proved that this property belongs to the geometrical magnitude represented by the general equation of the first degree

$$Ax+By+Cz+D=0,$$

it will follow that that magnitude is a plane.

Let $x = az + \alpha, \qquad y = bz + \beta,$

be the equations of a straight line, and suppose $x'$, $y'$, $z'$, to designate the co-ordinates of a point common to the line and surface. These co-ordinates will then satisfy the equations of the line and surface, and we shall have

$$x' = az' + \alpha, \qquad y' = bz' + \beta.$$

Also $Ax' + By' + Cz' + D = 0.$

Eliminating $x'$, $y'$, from these three equations, and we obtain

$$(Aa + Bb + C)z' + A\alpha + B\beta + D = 0,$$

which is the equation of condition that establishes a common point between the line and given surface.

If we suppose $x''$, $y''$, $z''$, to designate the co-ordinates of a second point common to the line and surface, we shall find

$$(Aa + Bb + C)z'' + A\alpha + B\beta + D = 0,$$

for the equation of condition.

Now, since these two equations of condition are true, whatever be the values of $z'$ and $z''$, it follows (Alg. Art. 208) that,

$$Aa + Bb + C = 0, \quad \text{and} \quad A\alpha + B\beta + D = 0;$$

and these are the equations of condition which cause two points of the straight line to be common with the surface.

Let us now take a third point of the straight line, and designate its co-ordinates by $x'''$, $y'''$, $z'''$.

If the co-ordinates $x'''$, $y'''$, $z'''$, will satisfy the equation of the surface, this third point will also be on the surface, But the equation of condition which will fix it in the surface, is

$$(Aa + Bb + C)z''' + A\alpha + B\beta + D = 0;$$

and this equation will be satisfied for any value of $z'''$, by virtue of the previous conditions.

Hence, if a straight line have two points in common with the surface represented by the general equation of the first degree between three variables; it will coincide with the surface, and consequently that surface is a plane.

## PROPOSITION IX. PROBLEM.

*To find the equation of a plane which shall pass through three given points.*

Let $x', y'\ z', \quad x'', y'', z'', \quad x''', y''', z'''$

be the co-ordinates of the given points.

The equation of the plane will be of the form

$$Ax + By + Cz + D = 0,$$

and it is required to find the values of $A$, $B$, and $C$, in terms of the co-ordinates of the given points.

Since the plane must pass through the three points, we shall have

$$Ax' + By' + Cz' + D = 0,$$

$$Ax'' + By'' + Cz'' + D = 0,$$

$$Ax''' + By''' + Cz''' + D = 0.$$

and placing the equations under the form

$$\frac{A}{D}x' + \frac{B}{D}y' + \frac{C}{D}z' + 1 = 0,$$

$$\frac{A}{D}x'' + \frac{B}{D}y'' + \frac{C}{D}z'' + 1 = 0$$

$$\frac{A}{D}x''' + \frac{B}{D}y''' + \frac{C}{D}z''' + 1 = 0.$$

We can readily find the values of $\frac{A}{D}$, $\frac{B}{D}$, and $\frac{C}{D}$ in terms of the given co-ordinates. We find

$$A = A'D, \qquad B = B'D, \qquad C = C'D,$$

in which $A'$, $B'$, $C'$, are functions of the co-ordinates of the given points.

Substituting these values in the equation of the plane, and dividing by $D$, we obtain

$$A'x + B'y + C'z + 1 = 0,$$

for the equation of the plane passing through the three given points.

*Scholium.* If one of the points be at the origin of co-ordinates, the equation of the plane will take the form

$$A''x + B''y + C''z = 0.$$

PROPOSITION X. PROBLEM.

*To find the equations of the intersection of two planes.*

Let
$$Ax + By + Cz + D = 0,$$
$$A'x + B'y + C'z + D' = 0,$$

be the equations of the two planes.

If the given planes intersect, the co-ordinates of their line of intersection will satisfy at the same time the equations of both planes.

Combining the two equations, and eliminating $z$, we obtain

$$(AC' - A'C)x + (BC' - B'C)y + (DC' - D'C) = 0;$$

which is the equation of the projection, on the plane of $YX$, of the line in which the planes intersect each other.

We may find, in a similar manner, the equation of the projection of the intersection on the co-ordinate plane $YZ$, and also on the co ordinate plane $ZX$.

*Scholium.* The method, which has just been explained, of determining the intersection of two planes, may be applied to any two surfaces whatever. For, the co-ordinates of the line of intersection, which is common to the two surfaces, will satisfy, at the same time, the equations of both surfaces. Hence, if the equations be combined, under the supposition that the co-ordinates are equal to each other, and one of the variables be eliminated, the resulting equation will be the equation of the projection of the intersection on one of the co-ordinate planes; and by eliminating each of the other variables in succession, the equations of the projections on the two other co-ordinate planes may be determined.

### PROPOSITION XI. PROBLEM.

*To find the conditions which will cause two planes to be parallel to each other.*

Let $Ax+By+Cz+D=0, \quad A'x+B'y+C'z+D'=0$

be the equations of the planes.

If these planes are parallel to each other, their traces on the co-ordinate planes will be respectively parallel.

We shall therefore have the following conditions (Prop. VII, Sch. 4),

$$\frac{A}{C}=\frac{A'}{C'}, \qquad \frac{B}{C}=\frac{B'}{C'}, \qquad \frac{A}{B}=\frac{A'}{B'},$$

in which we see, that either two of the conditions will give the third; and hence, if the traces be parallel on two of the co-ordinate planes, they will necessarily be parallel on the other, unless the planes were both parallel to the intersection of the co-ordinate planes, in which case, they might or might not be parallel.

*Scholium.* The relations which exist between the coefficients $A$, $B$, $C$, $A'$, $B'$, $C'$, when the two planes are parallel to each other, may be deduced from the equation (Prop. X)

$$(AC'-A'C)x+(BC'-B'C)y+(DC'-D'C)=0,$$

of the intersection of two planes.

If this equation be satisfied, the two planes will intersect each other, for the projection of their intersection can be constructed. If, however, this equation be not satisfied, that is, if it be rendered untrue, the projection of the intersection cannot exist: in this case the planes cannot intersect, and therefore they must be parallel.

If now, we make

$$AC'-A'C=0, \quad \text{and} \quad BC'-B'C=0,$$

the first two terms will reduce to 0; but these suppositions do not necessarily reduce the absolute term, $DC'-D'C$ to 0: hence, they will render the equation untrue, and therefore will cause the planes to be parallel.

We deduce from these equations of condition,

$$\frac{A}{C}=\frac{A'}{C'}, \qquad \frac{B}{C}=\frac{B'}{C'},$$

the same conditions as before found.

## PROPOSITION XII. PROBLEM.

*To find the conditions which will cause a straight line to be parallel to a plane.*

Let $\quad x=az+\alpha, \qquad y=bz+\beta,$

be the equations of the line, and

$$Ax+By+Cz+D=0,$$

the equation of the plane.

If a parallel line and a parallel plane be drawn through

the origin of co-ordinates, their equations will be

$$x = az, \qquad y = bz,$$

and

$$Ax + By + Cz = 0.$$

Now, if the first line be parallel to the first plane, the line through the origin of co-ordinates will coincide with the plane through the origin: and conversely, the condition which causes the line and plane drawn through the origin to coincide, will cause the first line and plane to be parallel to each other.

But the line and plane drawn through the origin will coincide (Prop. VIII) when

$$Aa + Bb + C = 0;$$

hence, this condition will cause the line and plane to be parallel to each other.

*Scholium.* This condition may also be deduced by supposing the line to intersect the plane.

For, if we suppose the line to pierce the plane, the co-ordinates of the common point will satisfy the equations of the line and plane. Combining them, and eliminating $x$ and $y$, we find

$$z = -\frac{A\alpha + B\beta + D}{Aa + Bb + C}.$$

But if we make

$$Aa + Bb + C = 0,$$

the value of $z$ will be infinite, and if the co-ordinates of the point of intersection are infinite, the line will be parallel to the plane.

PROPOSITION XIII. PROBLEM.

*To find the conditions which will cause a straight line to be perpendicular to a plane.*

Let $$x = az + \alpha, \qquad y = bz + \beta,$$

be the equations of the line, and

$$Ax + By + Cz + D = 0,$$

be the equation of the plane.

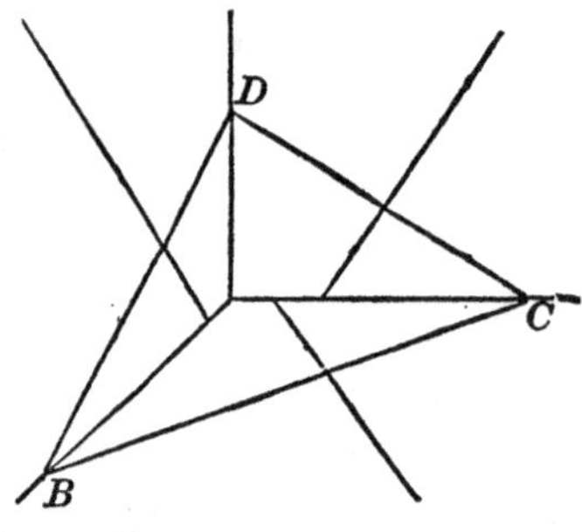


The equation of the trace $CD$ is

$$Ax + Cz + D = 0,$$

or $$x = -\frac{C}{A}z - \frac{D}{A}.$$

The equation of the trace $BD$ is

$$y = -\frac{C}{B}z - \frac{D}{B}.$$

But since the projections of the line must be respectively perpendicular to the traces of the plane (Prop. VII, Sch. 3), we shall have

$$a \times -\frac{C}{A} + 1 = 0, \qquad b \times -\frac{C}{B} + 1 = 0,$$

which gives

$$A = aC, \quad \text{and} \quad B = bC,$$

and these are the required conditions.

### PROPOSITION XIV. PROBLEM.

*To draw from a given point a line perpendicular to a given plane, and to find the length of the perpendicular.*

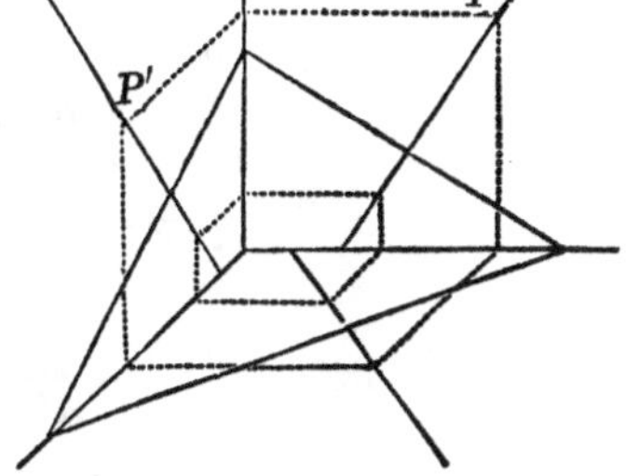


Let $(P, P')$ be the given point, of which the co-ordinates are $x'$, $y'$, $z'$, and let

$$Ax + By + Cz + D = 0,$$

be the equation of the given plane.

The equations of a line passing through the given point are,

$$x - x' = a(z - z'), \qquad y - y' = b(z - z').$$

If we make

$$D' = Ax' + By' + Cz' + D,$$

the equation of the given plane may be placed under the form

$$A(x - x') + B(y - y') + C(z - z') + D' = 0.$$

If we now suppose the line to intersect the plane, the co-ordinates of the common point, which we will designate by $x''$, $y''$, $z''$, will satisfy the equations of the line and plane, and by combining them, we shall have,

$$z'' - z' = -\frac{D'}{Aa + Bb + C},$$

$$x'' - x' = -\frac{aD'}{Aa + Bb + C},$$

$$y'' - y' = -\frac{bD'}{Aa + Bb + C}.$$

But the conditions which cause the line to be perpendicular to the plane, give

$$a = \frac{A}{C}, \quad \text{and} \quad b = \frac{B}{C}.$$

Substituting these values, and we obtain

$$z'' - z' = -\frac{CD'}{A^2 + B^2 + C^2},$$

$$x'' - x' = -\frac{AD'}{A^2 + B^2 + C^2},$$

$$y'' - y' = -\frac{BD'}{A^2 + B^2 + C^2}.$$

But the distance between the two points, of which the co-ordinates are $x''$, $y''$, $z''$, and $x'$, $y'$, $z'$, is,

$$\sqrt{(x'' - x')^2 + (y'' - y')^2 + (z'' - z')^2};$$

and if we designate this distance by $P$, we have

$$P = \frac{D'}{\sqrt{A^2 + B^2 + C^2}},$$

or by substituting for $D'$ its known value,

$$P = \frac{Ax' + By' + Cz' + D}{\sqrt{A^2 + B^2 + C^2}}.$$

*Scholium.* If the given point falls on the plane, its co-ordinates $x'$, $y'$, $z'$, will satisfy the equation of the plane. Under this supposition the numerator in the value of $P$ will reduce to 0; and hence, $P$ will become 0, as it should do when the point falls on the plane.

### PROPOSITION XV. PROBLEM.

*To find the angle included between two planes.*

Let
$$Ax + By + Cz + D = 0,$$
and
$$A'x + B'y + C'z + D' = 0,$$
be the equations of the two planes.

From any point in space draw two lines, respectively perpendicular to the given planes. The angle included between one of these lines, and the prolongation of the other, will be equal to the angle included between the planes.

The equations of the two perpendiculars will be of the form

$$x = az + \alpha, \qquad y = bz + \beta,$$

$$x = a'z + \alpha', \qquad y = b'z + \beta'.$$

But if these lines are respectively perpendicular to the planes, we shall have

$$A = aC, \qquad B = bC, \qquad A' = a'C', \qquad B' = b'C'.$$

If we now designate the angle included between the lines, which is the same as that included between the planes, by $V$ we shall have (Prop. V, Sch. 1),

$$\cos V = \frac{1 + aa' + bb'}{\sqrt{1 + a^2 + b^2}\,\sqrt{1 + a'^2 + b'^2}};$$

if we substitute for $a$, $b$, $a'$, $b'$, their values, we have

$$\cos V = \pm\frac{AA' + BB' + CC'}{\sqrt{A^2 + B^2 + C^2}\,\sqrt{A'^2 + B'^2 + C'^2}}.$$

*Scholium* 1. The sign $+$ or $-$ may be attributed to the radicals in the denominator, and hence, the value of $\cos V$ may be either positive or negative. The positive

19*

value will correspond to the acute, and the negative value to the obtuse angle.

*Scholium* 2. The value of cos $V$ is independent of $D$ and $D'$. Indeed, it ought not to depend upon them, since the distances cut off from the co-ordinate axes may be varied at pleasure, without affecting the inclination of the planes.

*Scholium* 3. If the two planes are perpendicular to each other, we shall have $\cos V = 0$, and

$$AA' + BB' + CC' = 0.$$

*Scholium* 4. If we now suppose one of the planes, the second for example, to coincide in succession with each of the co-ordinate planes, the corresponding values of $V$ will express the angles which the first plane makes with the co-ordinate planes, respectively.

If we suppose the second plane to coincide with the co-ordinate plane $YX$, of which the characteristic is $z = 0$, the equation of the plane will reduce to

$$A'x + B'y = 0;$$

and since this equation is true for all values of $x$ and $y$, we have

$$A' = 0, \qquad \text{and} \qquad B' = 0.$$

If we designate by $V'$ the angle formed by the planes, we shall have

$$\cos V' = \frac{C}{\sqrt{A^2 + B^2 + C^2}}.$$

If we designate by $V''$ and $V'''$ the angles which the first plane forms with the co-ordinate planes $ZX$, $YZ$, we shall find

$$\cos V'' = \frac{B}{\sqrt{A^2 + B^2 + C^2}}, \qquad \cos V''' = \frac{A}{\sqrt{A^2 + B^2 + C^2}}.$$

If we square the three values of cos $V'$, cos $V''$ cos, $V'''$, and add, we find

$$\cos^2 V' + \cos^2 V'' + \cos^2 V''' = 1;$$

that is, *the sum of the squares of the cosines of the three angles which a plane forms with the three co-ordinate planes, is equal to radius square or unity.*

*Scholium* 5. If we now suppose the first plane to coincide, in succession, with each of the co-ordinate planes, and designate by $U'$, $U''$, $U'''$, the angles formed by the second plane, with the co-ordinate planes, we shall find

$$\cos U' = \frac{C'}{\sqrt{A'^2 + B'^2 + C'^2}}, \qquad \cos U'' = \frac{B'}{\sqrt{A'^2 + B'^2 + C'^2}},$$

$$\cos U''' = \frac{A'}{\sqrt{A'^2 + B'^2 + C'^2}}.$$

If in the equation

$$\cos V = \frac{AA' + BB' + CC'}{\sqrt{A^2 + B^2 + C^2}\sqrt{A'^2 + B'^2 + C'^2}},$$

we substitute the values of $A$, $B$, $C$, $A'$, $B'$, $C'$, which may be found from the foregoing equations, we shall find

$$\cos V = \cos V' \cos U' + \cos V'' \cos U'' + \cos V''' \cos U''',$$

which last equation expresses the cosine of the angle included between two planes, in terms of the angles which the planes form with the co-ordinate planes.

## PROPOSITION XVI. PROBLEM.

***To find the angle included between a straight line and plane***

Let $\qquad x = az + \alpha, \qquad y = bz + \beta,$

be the equations of the line, and

$$Ax + By + Cz + D = 0,$$

the equation of the plane.

The angle included between a line and plane, is represented by the angle included between the line and its projection on the plane. If, therefore, from any point of the given line a perpendicular be drawn to the plane, the angle which it forms with the given line will be the complement of the required angle.

Let $$x = a'z + \alpha', \qquad y = b'z + \beta',$$

be the equations of the perpendicular. The cosine of the angle which it forms with the given line is expressed (Prop. V, Sch. 1) by

$$\frac{1 + aa' + bb'}{\sqrt{1 + a^2 + b^2}\sqrt{1 + a'^2 + b'^2}}.$$

But since the second line is perpendicular to the plane, we have

$$a' = \frac{A}{C}, \qquad b' = \frac{B}{C}.$$

Substituting these values, and representing the angle sought by $V$, we obtain

$$\sin V = \frac{Aa + Bb + C}{\sqrt{1 + a^2 + b^2}\sqrt{C^2 + B^2 + A^2}}.$$

*Scholium.* If we suppose the line to become parallel to the plane, we shall have $V = 0$, and consequently $\sin V = 0$: and hence,

$$Aa + Bb + C = 0,$$

the same condition as before found.

## *Examples on the last Book.*

1. What is the distance between two points, of which the co-ordinates are

$$x' = 5,\ y' = 5,\ z' = -3; \quad x'' = -1,\ y'' = 0,\ z'' = 5.$$

*Ans.* 11.18

2. The equations of the projections of a straight line on the co-ordinate planes $ZX$, $YZ$, are

$$x = z + 1, \qquad y = \frac{1}{2}z - 2,$$

required its equation on the plane $YX$.

*Ans.* $2y = x - 5$.

3. Required the equations of the three projections of a straight line which passes through the two points whose co-ordinates are,

$$x' = 2,\ y' = 1,\ z' = 0, \quad \text{and} \quad x'' = -3,\ y'' = 0,\ z'' = -1.$$

*Ans.* $x = 5z + 2$, $y = z + 1$, $5y = x + 3$.

4. Required the angle included between two lines, whose equations are

$$\left.\begin{array}{l} x = 3z + 5 \\ y = 5z + 3 \end{array}\right\} \text{of the 1st,}$$

and

$$\left.\begin{array}{l} x = z + 1 \\ y = 2z \end{array}\right\} \text{of the 2d.}$$

*Ans.* $14^\circ\ 58'$.

5. Required the angles which a straight line makes with the co-ordinate axes, its equations being

$$x = -2z + 1,$$

$$y = z + 3.$$

*Ans.* $\left\{\begin{array}{l} 144^\circ\ 44' \text{ with } X, \\ 65^\circ\ 54' \text{ with } Y, \\ 65^\circ\ 54' \text{ with } Z. \end{array}\right.$

6. Having given the equations of two straight lines,

$$\left.\begin{array}{l} x = 2z + 1 \\ y = 2z + 2 \end{array}\right\} \text{of the 1st,}$$

and

$$\left.\begin{array}{l} x = z + 5 \\ y = 4z + \beta' \end{array}\right\} \text{of the 2d,}$$

required the value of $\beta'$ so that the lines shall intersect each other, and to find the co-ordinates of the point of intersection.

$$Ans. \begin{cases} \beta' = -6, \\ x' = \;\; 9, \\ y' = \;\; 10, \\ z' = \;\; 4. \end{cases}$$

7. To find the equations of a line that shall pass through a point, of which the co-ordinates are $x' = -2$, $y' = 3$, $z' = 5$, and be perpendicular to the plane, of which the equation is

$$2x + 8y - z - 4 = 0.$$

$$Ans. \begin{cases} x = -2z + 8, \\ y = -8z + 43 \end{cases}$$

8. To find the equation of a plane which shall pass through the three points, whose co-ordinates are

$$x' = 1, \quad y' = -2, \quad z' = 2; \qquad x'' = 0, \quad y'' = 4, \quad z'' = -5,$$
$$x''' = -2, \quad y''' = 1, \quad z''' = 0.$$

$$Ans. \quad 9x + 19y + 15z - 1 = 0.$$

9. To find the equations of the intersection of two planes, of which the equations are

$$3x + 8y - 10z + 6 = 0, \quad \text{of the 1st,}$$

and

$$4x - 8y + z \quad + 1 = 0, \quad \text{of the 2d.}$$

$$Ans. \begin{cases} 7x - 9z + 7 = 0, \\ 56y - 43z + 21 = 0. \end{cases}$$

10. To find the traces of a plane whose equation is

$$x - 9y + 11z - 12 = 0.$$

11. To find the length of a line drawn from a point, whose co-ordinates are $x' = 2$, $y' = -3$, $z' = 0$, and perpendicular to a plane whose equation is

$$8x + 9y - z + 2 = 0.$$

$$Ans. \quad \frac{-9}{\sqrt{146}}$$

**12.** To find the angle included between two planes, whose equations are

$$5x - 7y + 3z + 1 = 0, \quad \text{of the 1st,}$$

and

$$2x + y - 3z = 0, \quad \text{of the 2d.}$$

*Ans.* 100° 08′.

**13.** To find the angle which the plane, whose equation is

$$5x - 7y + 3z + 1 = 0,$$

makes with the co-ordinate planes.

*Ans.* { 70° 46′ with the plane $XY$,
140° 12′ with the plane $ZX$,
56° 43′ with the plane $YZ$.

## *Transformation of Co-ordinates in Space.*

The transformation of co-ordinates in space, consists in finding the relations which exist between the co-ordinates of a series of points referred to two different systems of co-ordinate planes. We shall find, that in space, as well as in the plane (Bk. II, Prop. XIV, Sch. 2), the primitive co-ordinates of any point are expressed in terms of the new co-ordinates of the same point, the directions of the new axes, and the co-ordinates of the new origin.

### PROPOSITION XVII. PROBLEM.

*To find the formulas for passing from a system of co-ordinate planes at right angles to each other, to a new system of co-ordinate planes.*

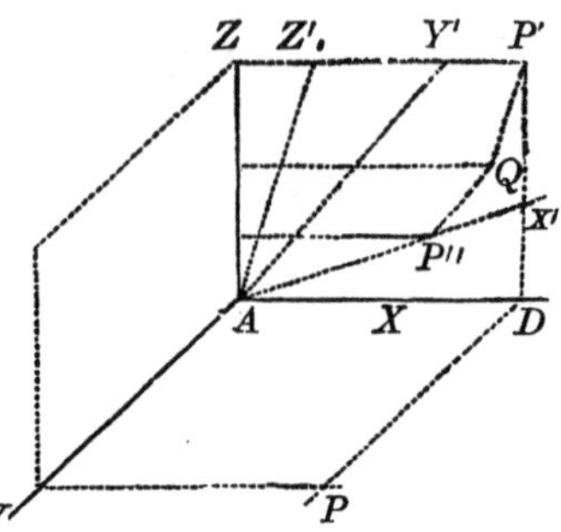


Let $(P, P')$ be any point in space, of which the co-ordinates are

$$AD=x, \quad PD=y, \quad P'D=z.$$

Through the origin $A$ draw three new axes, $AX'$, $AY'$, $AZ'$. Through the point $(P, P')$, draw a line parallel to the new axis of $Z'$, and through $Q$, the point in which the parallel pierces the plane $X'Y'$, draw $QP''$ parallel to the axis $Y'$. Then will $AP''=x'$, $QP''=y'$, $P'Q=z'$, be the co-ordinates of the point $(P, P')$, referred to the new system of co-ordinate planes.

Through the three points, $P''$, $Q$, $P'$, let three planes be passed parallel to the plane $YX$. It is now evident that the distance from the co-ordinate plane $YX$ to the first parallel plane will be the projection of $AP''$ or $x'$ on the axis of $Z$; the distance between the first and second planes, the projection of $y'$ on the axis of $Z$, and the distance between the second and third, the projection of $z'$ on the axis of $Z$. It is also plain, that the sum of the three distances will be equal to the co-ordinate $z$ of the point $(P, P')$, referred to the primitive system of co-ordinates.

If we designate by $Z$, $Z'$, $Z''$, the three angles which the new axes $X'$, $Y'$, $Z'$, make respectively with the primitive axis of $Z$, we shall have

$$z = x' \cos Z + y' \cos Z' + z' \cos Z''.$$

If we designate by $X$, $X'$, $X''$ the three angles which the new axes make, respectively, with the primitive axis of $X$, and by $Y$, $Y'$, $Y''$, the three angles which the new axes make, respectively, with the primitive axis of $Y$, we may find, by a course of reasoning entirely similar to the above,

$$\left.\begin{aligned} x &= x' \cos X + y' \cos X' + z' \cos X'' \\ y &= x' \cos Y + y' \cos Y' + z' \cos Y'' \\ z &= x' \cos Z + y' \cos Z' + z' \cos Z'' \end{aligned}\right\} \quad (1).$$

*Scholium* 1. If we suppose, at the same time, the origin to be changed, and designate the co-ordinates of the new origin by $a$, $b$, $c$, the equations will become

$$x = a + x' \cos X + y' \cos X' + z' \cos X'',$$

$$y = b + x' \cos Y + y' \cos Y' + z' \cos Y'',$$

$$z = c + x' \cos Z + y' \cos Z' + z' \cos Z''.$$

*Scholium* 2. If the new axes are parallel to the primitive axes, the equations for transformation will become

$$x = a + x', \qquad y = b + y', \qquad z = c + z'.$$

*Scholium* 3. We may consider each of the new axes as forming three angles with the primitive axes of $X$, $Y$, and $Z$; and since these axes are at right angles to each other, we shall have (Prop. I, Sch. 4),

$$\left.\begin{aligned} \cos^2 X + \cos^2 Y + \cos^2 Z &= 1 \\ \cos^2 X' + \cos^2 Y' + \cos^2 Z' &= 1 \\ \cos^2 X'' + \cos^2 Y'' + \cos^2 Z'' &= 1 \end{aligned}\right\} \quad (2).$$

*Scholium* 4. The angles which the new axes form with each other are yet undetermined. Let us designate by

$V$ the angle formed by $X'$ and $Y'$,

$U$ the angle formed by $Y'$ and $Z'$,

$W$ the angle formed by $Z'$ and $X'$

Since the angles which the new axes form with the primitive axes have already been designated, we shall have (Prop. V),

$$\left.\begin{aligned} \cos V &= \cos X \cos X' + \cos Y \cos Y' + \cos Z \cos Z' \\ \cos U &= \cos X' \cos X'' + \cos Y' \cos Y'' + \cos Z' \cos Z'' \\ \cos W &= \cos X \cos X'' + \cos Y \cos Y'' + \cos Z \cos Z'' \end{aligned}\right\} (3).$$

*Scholium* 5. In passing from one system of co-ordinate planes to another we introduce the angles which each new axis makes with the primitive axes. The formulas for transformation will therefore contain nine arbitrary constants, and if the origin be changed at the same time, they will contain twelve.

Now, if the angles which the new axes make with the primitive axes are known, the coefficients of $x'$, $y'$, $z'$, in equations (1), will be known, and the angles which the new axes form with each other may be found from equations (3).

If, however, the new axes are to be so chosen as to fulfil particular conditions, these conditions will be expressed by assigning suitable values to the co-ordinates of the new origin, and the angles which the new axes form with the primitive Suppose, for example, it were required that the new axes should be at right angles to each other: we should then have

$$\cos V = 0, \qquad \cos U = 0, \qquad \cos W = 0,$$

and consequently

$$\cos V = \cos X \cos X' + \cos Y \cos Y' + \cos Z \cos Z' = 0$$
$$\cos U = \cos X' \cos X'' + \cos Y' \cos Y'' + \cos Z' \cos Z'' = 0$$
$$\cos W = \cos X \cos X'' + \cos Y \cos Y'' + \cos Z \cos Z'' = 0$$

*Scholium* 6. Let us suppose it were required that the new axes $X'$, $Y'$, should fall in the primitive co-ordinate plane $YX$, and that the new axis of $Z'$ should coincide with the primitive axis of $Z$. These conditions will give

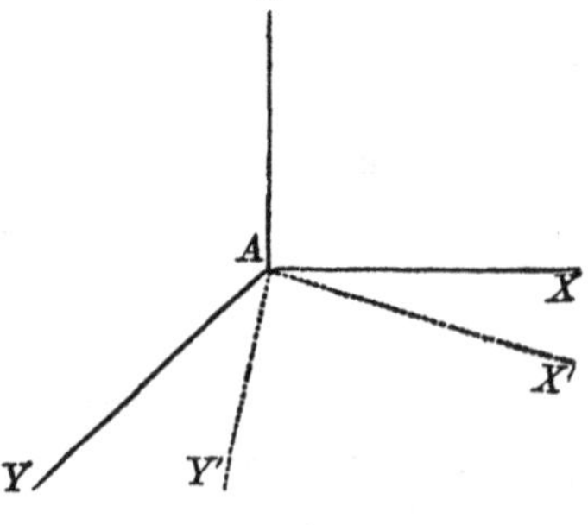


$$ZAZ' = 0, \qquad ZAX' = 90°, \qquad ZAY' = 90°,$$

$$U = 90°, \qquad \text{and} \qquad W = 90°;$$

and hence, $\cos Z'' = 1$, $\cos X'' = 0$, $\cos Y'' = 0$:

also, $\cos U = 0$, and $\cos W = 0$.

These values being substituted in the two last of equations (3), they become

$$\cos Z' = 0, \qquad \cos Z = 0.$$

If these several values be substituted in equations (2), they become

$$\cos^2 X + \cos^2 Y = 1, \qquad \cos^2 X' + \cos^2 Y' = 1;$$

and consequently we have

$$\cos Y = \sin X, \quad \text{and} \quad \cos Y' = \sin X'.$$

If these values be substituted in equations (1), they become

$$x = x' \cos X + y' \cos X', \qquad y = x' \sin X + y' \sin X',$$

which are the formulas for passing from rectangular to oblique co-ordinates (Bk. II, Prop. XI).

## *Of Polar Co-ordinates in Space.*

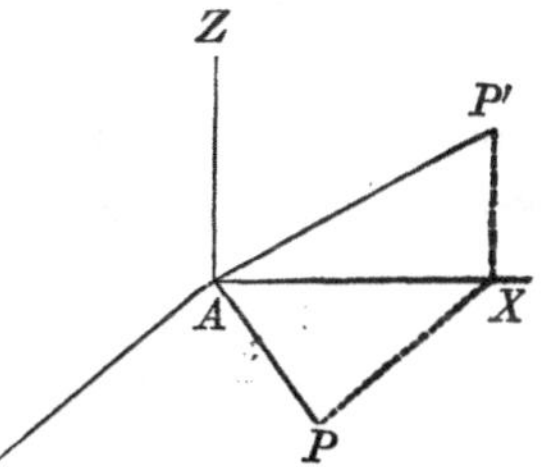


Let $(P, P')$ be any point in space, and $(AP, AP')$ a line passing through it and the origin of co-ordinates. This line is called a *radius-vector.*

Let us designate the radius-vector by $r$, and its projection $AP$, on the co-ordinate plane $YX$, by $r'$. Let us also designate the angle which the radius-vector forms with the co ordinate plane $YX$ by $u$, and the angle which its projection $AP$ forms with the axis of $X$ by $v$. We shall then have

$$x = r' \cos v \qquad y = r' \sin v, \qquad z = r \sin u;$$

also, $r' = r \cos u$:

hence we have

$$x = r\cos v\cos u, \qquad y = r\sin v\cos u, \qquad z = r\sin u,$$

these formulas may be applied to every point in space by attributing suitable values to $r$, $v$, and $u$.

In the equations

$$x = az, \qquad y = bz,$$

of a straight line passing through the origin of co-ordinates, $a$ and $b$ represent the tangents of the angles, which the projections on the co-ordinate planes $ZX$ and $YZ$, form with the axis of $Z$, and the values of the tangents are expressed by

$$\frac{x}{z} = a \quad \text{and} \quad \frac{y}{z} = b.$$

If we divide the first and second of the last equations by the third, we shall obtain

$$\frac{x}{z} = \frac{\cos v\cos u}{\sin u} \quad \text{and} \quad \frac{y}{z} = \frac{\sin v\cos u}{\sin u}:$$

hence, $$a = \frac{\cos v\cos u}{\sin u}, \qquad b = \frac{\sin v\cos u}{\sin u},$$

and therefore the values of $a$ and $b$ may be found when the values of $v$ and $u$ are known.

# BOOK IX.

## *Of Surfaces of the Second Order*

1. The equation of a surface is an equation expressing the relation between the co-ordinates of every point of the surface.

It has been shown (Bk. II, Prop. II), that every equation of the first degree between two variables, represents a straight line; and in (Bk. VII), that every equation of the second degree between two variables represents a curve.

It has also been shown (Bk. VIII, Prop. VIII), that every equation of the first degree between three variables represents a plane, and analogy would lead us to infer what will hereafter be rigorously proved, viz: that every equation of the second degree between three variables represents a curved surface.

2. Surfaces, like lines, are classed according to the degree of their equations. The plane, whose equation is of the first degree, is a surface of the *first order*, and every surface whose equation is of the second degree, is a surface of the *second order*.

3. The equation of a surface is its analytical representation, and although the equation determines the surface, yet it does not readily present to the mind its form, its dimensions, and its limits. To enable us to conceive of these, we intersect the surface by a system of planes, parallel, for example, to the co-ordinate planes. If then, we combine the equations of these planes with the equation of the surface, the resulting equations will represent the curves in which the planes

intersect the surface. These curves will show the form, the dimensions, and the limits of the surface.

4. To give a single example let us take the equation

$$x^2 + y^2 + z^2 = R^2.$$

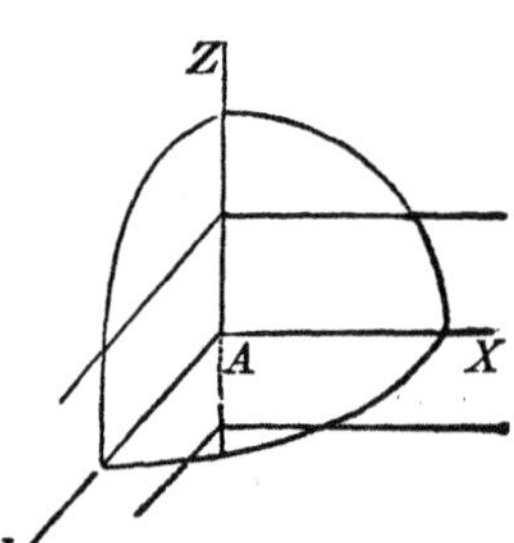


Let us intersect the surface represented by this equation by a plane parallel to $YX$, and at a distance from it equal to $c$. e equations of the plane will be

$$z = \pm c.$$

Combining this with the equation of the surface, and we shall have

$$x^2 + y^2 = R^2 - c^2,$$

which is the equation of the projection of the intersection on the co-ordinate plane $YX$ (Bk. VIII, Prop. II, Sch. 5). This equation represents the circumference of a circle whose centre is the origin of co-ordinates and radius $\sqrt{R^2 - c^2}$. The radius will be real for all values of $c$ less than $R$, whether $c$ be plus or minus. It is nothing when $c$ is equal to $R$, and imaginary when $c$ is greater than $R$. Thus, in the first case, the intersection will be the circumference of a circle, in the second case it will be a point, and in the third it will be an imaginary curve; or in other words, the plane will not intersect the surface.

Since the proposed equation is symmetrical with respect to the three variables $x$, $y$, and $z$, we may obtain similar results by intersecting the surface by planes parallel to the co-ordinate planes $YZ$ and $ZX$. The co-ordinate planes intersect the surface in circles whose equations are

$$x^2 + y^2 = R^2, \qquad x^2 + z^2 = R^2, \qquad y^2 + z^2 = R^2.$$

These results indicate that the surface whose equation is

$$x^2 + y^2 + z^2 = R^2,$$

is the surface of a sphere; but in order to prove it rigorously, it would be necessary to show that every secant plane would intersect the surface in the circumference of a circle.

## *Of the Surface of the Sphere.*

5. The general equation of the surface of a sphere will express the condition, that every point of the surface is equally distant from the centre.

6. Let us designate the co-ordinates of a fixed point by $x'$, $y'$, $z'$, and the co-ordinates of a point whose position may be changed at pleasure, by $x$, $y$, and $z$. If it be required that the second point shall be at a given distance from the first, and if we designate that distance by $R$, we shall have (Bk. VIII, Prop. I),

$$(x - x')^2 + (y - y')^2 + (z - z')^2 = R^2,$$

which is the general equation of the surface of a sphere, and in which, $x'$, $y'$, $z'$, are the co-ordinates of the centre, and $x$, $y$, and $z$, the general co-ordinates of the surface.

7. If the centre be placed at the origin of co-ordinates, the equation will reduce to

$$x^2 + y^2 + z^2 = R^2.$$

8. Let it be now required to pass a plane tangent to a sphere at a given point of the surface.

Let $x''$, $y''$, $z''$, be the co-ordinates of the given point, and

$$(x - x')^2 + (y - y')^2 + (z - z')^2 = R^2,$$

the equation of the surface.

Since the point of tangency is on the surface, we shall have

$$(x''-x')^2+(y''-y')^2+(z''-z')^2=R^2,$$

which may be placed under the form

$$(x''-x')(x''-x')+(y''-y')(y''-y')+(z''-z')(z''-z')=R^2. \quad (1)$$

The equation of a plane passing through the point whose co-ordinates are $x''$, $y''$, $z''$, is

$$A(x-x'')+B(y-y'')+C(z-z'')=0. \quad (2)$$

If we now draw a line through the centre of the sphere and the point of contact, its equation will be of the form (Bk VIII, Prop. III),

$$x-x''=\frac{x''-x'}{z''-z'}(z-z''), \qquad y-y''=\frac{y''-y'}{z''-z'}(z-z'').$$

But if the plane is tangent to the sphere the radius will be perpendicular to it at the point of contact (Geom. Bk. VIII, Prop VIII): hence, we have (Bk. VIII, Prop. XIII),

$$A=aC \qquad \text{and} \qquad B=bC,$$

that is,

$$A=\frac{x''-x'}{z''-z'}C, \qquad B=\frac{y''-y'}{z''-z'}C.$$

Substituting these values in the equation of the plane (2) and dividing by $C$, we obtain

$$(x''-x')(x-x'')+(y''-y')(y-y'')+(z''-z')(z-z'')=0, \quad (3)$$

which is the equation of a plane perpendicular to a line passing through the points whose co-ordinates are

$$x',\ y',\ z', \qquad\qquad x'',\ y'',\ z''.$$

But that one of these points may be the centre of a sphere and the other lie upon its surface, it will be necessary to combine the last equation with equation (1). Adding these equations, we obtain

$$(x''-x')(x-x')+(y''-y')(y-y')+(z''-z')(z-z')=R^2,$$

which is the equation of the tangent plane.

9. If the centre of the sphere be placed at the origin of co-ordinates, we shall have

$$x'=0, \quad y'=0, \quad \text{and} \quad z'=0,$$

and the equation of the tangent plane will become

$$x\,x''+y\,y''+z\,z''=R^2$$

## *Of Cylindrical Surfaces.*

10. A cylindrical surface may be generated or described by a straight line moving on a fixed curve and continuing parallel to itself in all its positions. The moving line is called the *generatrix*, and the fixed curve the *directrix* of the surface.

11. If the directrix be a curve of single curvature, let its plane be taken for one of the co-ordinate planes.

Let us suppose, for example, that the co-ordinate plane $YX$ contains the directrix. The equation of the directrix will then be expressed in terms of the variables $x$ and $y$ and constant quantities. When the directrix is a known curve the equation expressing the relations between $x$ and $y$ will be known.

12. In order to obtain the general equation of a cylindrical surface, that is, an equation which may be applied to a cylinder having any plane curve for its directrix, we place the equation of the directrix under the form

$$F(x, y)=0, \quad (1)$$

which is read, function of $x$, $y$, equal to 0; and the expression implies that $x$ and $y$ are dependent on each other, and may be made to represent the co-ordinates of any curve.

The equation of the generatrix will be of the form

$$x = az + \alpha, \qquad y = bz + \beta,$$

from which we shall have

$$\alpha = x - az, \quad \text{and} \quad \beta = y - bz. \quad (2)$$

Now, since the generatrix continues parallel to itself, $a$ and $b$ will have the same values for every position which it assumes: but the values of $\alpha$ and $\beta$ will continually change as the generating line moves around the directrix.

But since $\alpha$ and $\beta$ are the co-ordinates of the points in which the generatrix pierces the co-ordinate plane $YX$, (Bk. VIII, Prop. II, Sch. 4), and since these points must also lie on the directrix of the surface, the values of $\alpha$ and $\beta$, in equations (2), being substituted for $x$ and $y$, in equation (1), must satisfy that equation. Making these substitutions we have

$$F(x - az, y - bz) = 0,$$

and this is the general equation of the surface of a cylinder.

13. To apply this general equation to a particular case, let us suppose the directrix to be a circle having its centre at the origin of co-ordinates, and let us also suppose the generatrix of the surface to be oblique to the co-ordinate plane $YX$.

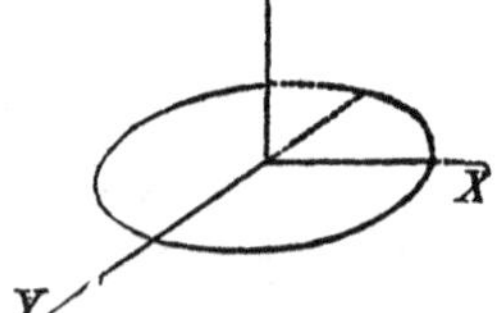


The equation of the directrix, which is of the form

$$F(x, y) = 0,$$

will become, by designating the radius by $r$,

$$F(x, y) = x^2 + y^2 - r^2 = 0.$$

Substituting for $x$ and $y$ in this equation, the values of $\alpha$ and $\beta$, we have

$$(x - az)^2 + (y - bz)^2 = r^2,$$

which is the equation of the surface of an oblique cylinder with a circular base.

14. If the base of the cylinder be an ellipse, of which the equation is

$$A^2y^2 + B^2x^2 = A^2B^2,$$

we shall have

$$F(x, y) = A^2y^2 + B^2x^2 - A^2B^2 = 0.$$

And the equation of the surface will become

$$A^2(y - bz)^2 + B^2(x - az)^2 = A^2B^2,$$

which is the equation of a cylindrical surface having an elliptical base.

15. If the generatrix becomes perpendicular to the co-ordinate plane $YX$, we shall have

$$a = 0, \qquad \text{and} \qquad b = 0,$$

the axis of the cylinder will then coincide with the axis of $Z$ and the equation of the surface will become

$$x^2 + y^2 = r^2,$$

for the circular base, and

$$A^2y^2 + B^2x^2 = A^2B^2,$$

for the elliptical base.

These last equations are the same as those for the corresponding directrices; but there is this difference, that for the equations of the directrices we have the further condition of

$$z = 0,$$

while in the equations of the surfaces $z$ may have any value whatever; which shows, that every section of the surface by a plane parallel to the co-ordinate plane $YX$, is equal to the directrix.

## *Equation of the Surface of the Cone.—Of the Sections of the Cone.*

16. A conical surface is that which may be described by the movement of a straight line constantly passing through a fixed point in space, and touching a given curve.

The fixed point is called the *vertex* of the cone; the curve, the *directrix*, and the moving line, the *generatrix* of the surface.

If the generatrix be prolonged in both directions, there will be two parts of the surface of which the vertex will be a common point: each of these parts is called a *nappe* of the cone.

17. Suppose the directrix to be placed in the co-ordinate plane $YX$, and to be represented by the equation

$$F(x, y) = 0.$$

Let us designate the co-ordinates of the fixed point, through which the generatrix is to pass, by $x'$, $y'$, $z'$: the equations of the generatrix will then be of the form

$$x - x' = a(z - z'), \qquad y - y' = b(z - z'), \quad (1)$$

from which we have

$$x = az + (x' - az'), \qquad y = bz + (y' - bz'), \quad (2)$$

and

$$a = \frac{x - x'}{z - z'}, \qquad b = \frac{y - y'}{z - z'}. \quad (3)$$

The terms $(x' - az')$ and $(y' - bz')$, of equations (2), are the co-ordinates of the points in which the generatrix pierces the co-ordinate plane $YX$: hence, the equation of the surface will become

$$F(x' - az', y' - bz') = 0, \quad (4)$$

in which $a$ and $b$ are constants for one position of the generatrix, but vary as the generatrix passes from one position to another.

If it be required to find the equation of the surface in terms of the co-ordinates of its different points, we must substitute in the last equation, the values of $a$ and $b$ found in equations (3): the equation of the surface then becomes

$$F\left(\frac{x'z - xz'}{z - z'}, \frac{y'z - yz'}{z - z'}\right) = 0. \quad (5)$$

18. If it be required to find the equation of the surface of a right cone with an elliptical base, and whose axis shall coincide with the axis of $Z$, we shall have, for the equation of the directrix,

$$A^2y^2 + B^2x^2 = A^2B^2.$$

We shall also have

$$x' = 0, \qquad y' = 0, \qquad z' = c$$

in which $c$ designates the distance from the origin of co-ordinates to the vertex of the cone. These values being substituted in equation (5), we obtain

$$F\left(\frac{-cx}{z - c}, \frac{-cy}{z - c}\right) = 0.$$

Substituting these values for $x$ and $y$ in the equation of the directrix, and we have

$$A^2y^2 + B^2x^2 = \frac{(z-c)^2}{c^2}.A^2B^2,$$

which is the equation of a right cone with an elliptical base.

19. If it were required to find the equation of an oblique cone with a circular base, the centre of the circle being at the origin of co-ordinates, the equation of the directrix would be

$$x^2 + y^2 = r^2.$$

Substituting for $x$ and $y$ the values

$$\frac{x'z - xz'}{z - z'}, \quad \frac{y'z - yz'}{z - z'},$$

which correspond to them in equation (5), and we shall obtain

$$(x'z - xz')^2 + (y'z - yz')^2 = r^2(z - z')^2,$$

which is the equation of an oblique cone with a circular base

20. The oblique cone, whose equation has just been found, will become a right cone with a circular base, if the vertex be placed on the axis of $Z$; we shall then have

$$x' = 0, \quad y' = 0, \quad z' = c;$$

and hence, the last equation will become

$$(x^2 + y^2)c^2 = r^2(z - c)^2,$$

which is the equation of a right cone with a circular base.

If we make

$$AC = c, \qquad AB = r,$$

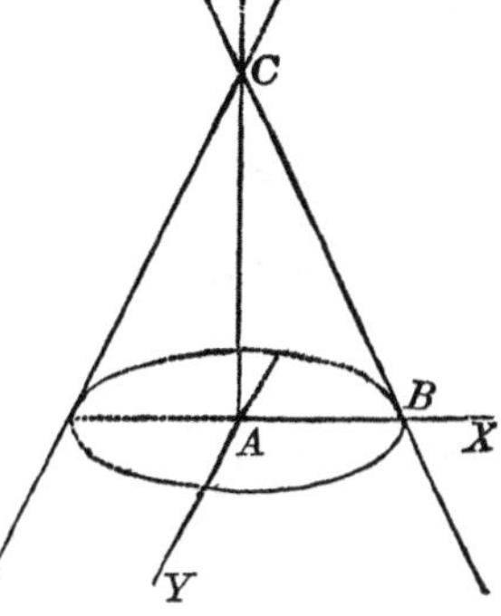


and designate the angle $ABC$, which the generatrix makes with the co-ordinate plane $YX$, by $v$, we shall have

$$\text{tang } v = \frac{c}{r},$$

and the equation of the surface will become

$$(y^2 + x^2) \text{ tang}^2 v = (z - c)^2.$$

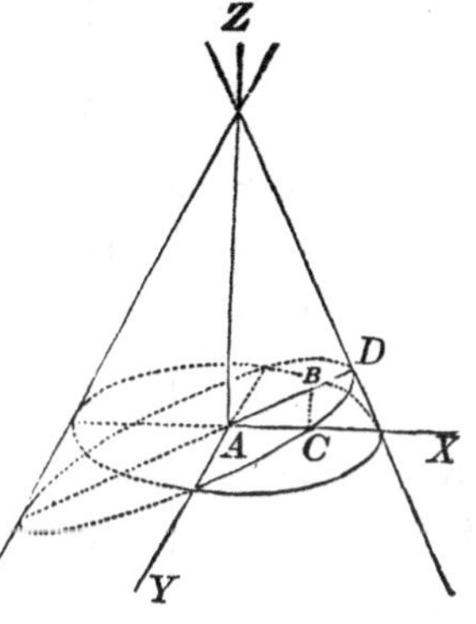


21. Let the surface of this cone be now intersected by a plane passing through the axis of $Y$, and consequently perpendicular to the co-ordinate plane $ZX$; and designate by $u$ the angle $DAX$, which the secant plane makes with the co-ordinate plane $YX$. The equation of this plane will be the same as that of its trace $AD$ (Bk. VIII, Prop. II, Sch. 3): that is,

$$z = x \text{ tang } u.$$

If we combine this equation with the equation of the surface, and eliminate $z$, we shall obtain the equation of the projection of the curve of intersection on the co-ordinate plane $YX$. It is, however, better to discuss the curve in its own plane, and for this purpose we will refer it to the two axes, $AY$, $AD$, which are in the plane of the curve, and at right angles to each other.

If we designate the co-ordinates of any point referred to

these axes, the one for example which is projected at $B$, by $x'$, $y'$, we shall have

$$AC = x = x' \cos u, \qquad BC = z = x' \sin u$$

and since the axis of $Y$ is not changed,

$$y = y'.$$

If we substitute these values in the equation of the surface of the cone, we shall obtain for the equation of the intersection, after reducing,

$$y'^2 \operatorname{tang}^2 v + x'^2 \cos^2 u (\operatorname{tang}^2 v - \operatorname{tang}^2 u) + 2cx' \sin u = c^2;$$

or omitting the accents,

$$y^2 \operatorname{tang}^2 v + x^2 \cos^2 u (\operatorname{tang}^2 v - \operatorname{tang}^2 u) + 2cx \sin u = c^2.$$

Since this equation is of the second degree, every curve of intersection will belong to one of three classes (Bk. VII Art. 10), which are characterized by

$$B^2 - 4AC < 0, \quad B^2 - 4AC = 0, \quad B^2 - 4AC > 0.$$

By comparing the equation of the curve of intersection with the general equation of the second degree, we find

$$B = 0, \quad A = \operatorname{tang}^2 v, \quad C = \cos^2 u (\operatorname{tang}^2 v - \operatorname{tang}^2 u).$$

Now we shall have

$$-4AC < 0,$$

when $A$ and $C$ have the same sign; and since $\operatorname{tang}^2 v$ and $\cos^2 u$ are positive, $A$ and $C$ will have the same sign when

$$\operatorname{tang} v > \operatorname{tang} u,$$

or

$$u < v;$$

and when this is the case, the curve of intersection will be an ellipse.

We shall also have

$$B^2 - 4AC = 0,$$

when $\tang v = \tang u$, or when $v = u$, in which case the curve of intersection will be a parabola.

If $\tang u > \tang v$, $A$ and $C$ will have different signs, and the curve of intersection will be an hyperbola.

22. In order to obtain the forms and classes of the curves which result from the intersection of the cone and plane, it might, at first, seem necessary to cause the angle $u$ to vary from 0 to 360°. But since the surface of the cone is symmetrical with respect to its axis, it is plain that all the varieties will be obtained by varying $u$ from 0 to 90°.

23. Let us then resume the equation of intersection,

$$y^2 \tang^2 v + x^2 \cos^2 u (\tang^2 v - \tang^2 u) + 2cx \sin u = c^2,$$

and begin the discussion of it by supposing

$$u = 0,$$

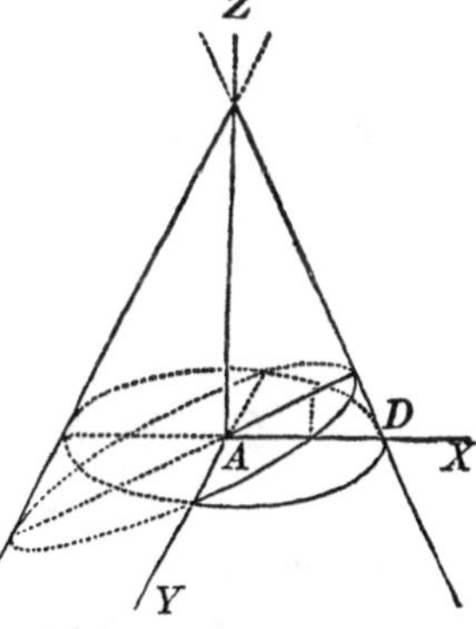


which will cause the secant plane to coincide with the co-ordinate plane $YX$. The equation of the curve will then become

$$x^2 + y^2 = \frac{c^2}{\tang^2 v};$$

hence, the curve is the circumference of a circle, of which $A$ is the centre, and $AD$ equal to $\dfrac{c}{\tang v}$ the radius.

24. If we now suppose $u$ to increase, the curve of intersection will be an ellipse so long as $u < v$: that is, *if a right cone with a circular base be intersected by a plane making with the base of the cone an angle less than the angle*

*formed by the element and base, all the elements of the same nappe will be intersected, and the curve of intersection will be an ellipse.*

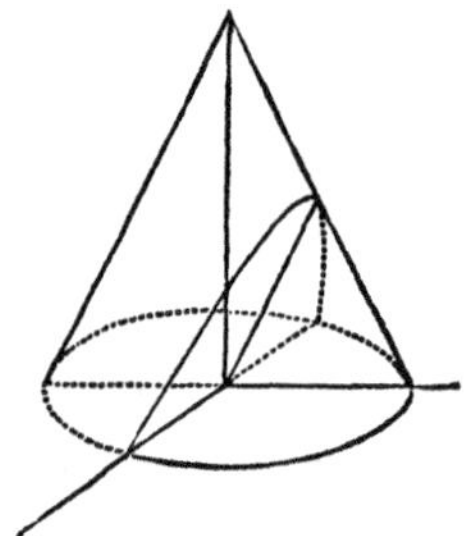

25. When $u$ becomes equal to $v$, the cutting plane becomes parallel to an element of the cone: hence, *if a right cone with a circular base be intersected by a plane parallel to one of the elements, the curve of intersection will be a parabola.*

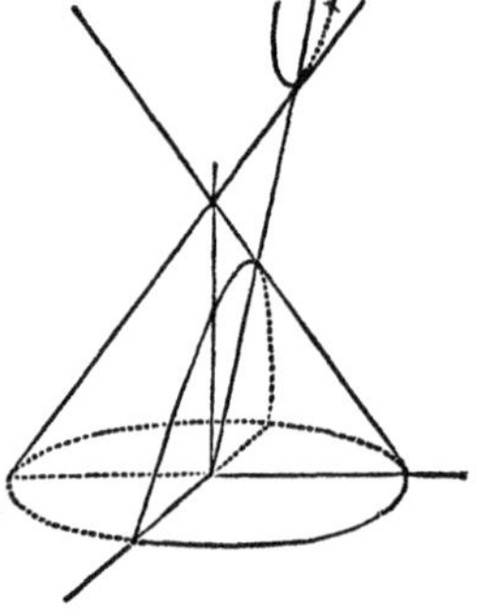

26. When $u$ becomes greater than $v$, the cutting plane will intersect both nappes of the cone: hence, *if a right cone with a circular base be intersected by a plane making with the base of the cone an angle greater than the angle formed by the element and base, both nappes of the cone will be intersected, and the curve of intersection will be an hyperbola.*

## *Of the Surfaces of Revolution.*

27. Every surface which can be generated by the revolution of a line about a fixed axis, is called a *surface of revolution.*

The revolving line is called the *generatrix;* and the line about which it revolves is called the *axis of the surface,* or the *axis of revolution.* The section made by a plane passing through the axis, is called a *meridian section,* or a meridian curve when the surface is of double curvature

28. It is plain, from the definition of a surface of revolution, that every point of the generatrix will describe the circumference of a circle, the centre of which is in the axis of revolution.

29. If the generating curve be of single curvature, we may assume its plane for one of the co-ordinate planes; and if the axis about which it is revolved be a straight line, one of the co-ordinate axes may be taken to coincide with this line.

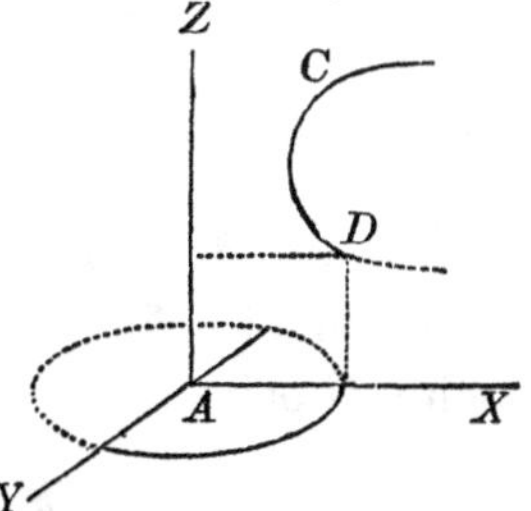


30. Let $DC$ be any curve in the co-ordinate plane $ZX$, and let it be revolved around the axis of $Z$: it is required to determine the equation of the surface which it will describe.

If we designate the abscissa of any point of the generatrix, as $D$, by $r$, and the ordinate by $z$, the equation of the generatrix may be written under the form

$$r = F(z):$$

the value of $r$ in terms of $z$ and constants, may always be found when the equation of the generatrix is known.

We have now to express, analytically, the conditions which will cause any point of the generatrix, as $D$, to describe the circumference of a circle around the axis of $Z$. To do this, we have only to consider, that the circumference described by any point of the generatrix, as $D$, will be projected on the co-ordinate plane $YX$ into an equal circumference. If the co-ordinates of the points of this circumference be designated by $x$ and $y$, we shall have

$$r = \sqrt{x^2 + y^2},$$

If we now suppose $r$ to take all possible values which

will satisfy the equation

$$r = F(z),$$

and combine the two equations, we shall obtain

$$F(z) = \sqrt{x^2 + y^2},$$

which is the equation of the surface of revolution.

31. As a first example, let it be required to determine the equation of the surface described by the revolution of a straight line which intersects the axis of $Z$ at a given point.

If we designate the co-ordinates of the given point by

$$x' = 0, \qquad y' = 0, \qquad z' = c,$$

the equation of a line in the co-ordinate plane $ZX$, passing through this point will be

$$x = a(z - c),$$

and hence the equation of the generatrix will be

$$r = a(z - c),$$

and this equation shows the value of $r$ in terms of $z$, or the value of $F(z)$. Substituting this value in the general equation of the surface, and we obtain

$$a^2(z - c)^2 = x^2 + y^2,$$

or,

$$x^2 + y^2 = a^2(z - c)^2,$$

which is the equation of the surface described by the straight line. This surface is that of a right cone with a circular base, and its equation is the same as that found in (Art. 20), since the angle formed by the element and axis is the complement of the angle formed by the element and base and hence,

$$\text{tang } v = \frac{1}{a}.$$

If the vertex be placed at the origin of co-ordinates, we shall have $c = 0$, and the equation of the surface will become

$$x^2 + y^2 = a^2z^2.$$

32. If it be required to find the equation of the surface of a sphere, of which the centre shall be at the origin of co-ordinates, the equation of the generatrix will be

$$r^2 + z^2 = R^2,$$

or,
$$r = \sqrt{R^2 - z^2},$$

Substituting this value in the general equation (Art. 30), and we obtain

$$x^2 + y^2 + z^2 = R^2,$$

for the equation of the surface of the sphere.

33. The surface described by the revolution of an ellipse about either axis, is called an *ellipsoid of revolution*. It is also sometimes called a *spheroid*. It is called a *prolate spheroid* when the ellipse is revolved about the transverse axis, and an *oblate spheroid*, when it is revolved about the conjugate axis.

34. Let it be now required to find the equation of the surface of a prolate spheroid. If the transverse axis of the ellipse coincides with the axis of $Z$, the equation of the generatrix will be of the form

$$B^2z^2 + A^2r^2 = A^2B^2,$$

hence,
$$r = \sqrt{\frac{A^2B^2 - B^2z^2}{A^2}}.$$

Substituting this value in the general equation of the sur

face of revolution, and we obtain

$$B^2 z^2 + A^2 (x^2 + y^2) = A^2 B^2,$$

which is the equation of the surface of a prolate spheroid.

35. We should find, by a similar process the equation of the surface of the oblate spheroid, to be

$$A^2 z^2 + B^2 (x^2 + y^2) = A^2 B^2.$$

If in either of these equations we make $A = B$, we shall obtain

$$x^2 + y^2 + z^2 = R^2,$$

the equation of the surface of a sphere.

36. The surface described by the revolution of a hyper bola about its transverse axis, is called a *hyperboloid of revolution of two nappes.*

The surface described by the revolution of a hyperbola about its conjugate axis, is called a *hyperboloid of revolution of one nappe.*

37. The equation of the hyperbola referred to its centre and axes, is

$$A^2 y^2 - B^2 x^2 = - A^2 B^2,$$

and by revolving the curves about the transverse axis, we find

$$B^2 z^2 - A^2 (x^2 + y^2) = A^2 B^2,$$

for the equation of the surface described; and by revolving around the conjugate axis, the equation of the surface will be

$$A^2 z^2 - B^2 (x^2 + y^2) = - A^2 B^2.$$

If the asymptotes revolve around the transverse axis, they will describe the surface of a cone with two nappes. The surface of this cone will approach the surfaces of the nappes of the hyperboloid, and will become tangent to them at an infinite distance from the centre.

38. If a parabola be revolved around its principal axis, the surface described is called a *paraboloid of revolution.* The equation of the generatrix being of the form

$$r^2 = 2pz,$$

the equation of the surface will be

$$x^2 + y^2 = 2pz.$$

39. If the generatrix does not coincide with the co-ordinate plane of $ZX$, its equations will be of the form

$$x = F(z), \quad (1)$$

$$y = F'(z). \quad (2)$$

If the generatrix be revolved around the axis of $Z$, every point of it will describe the circumference of a circle of which the centre will be in the axis.

If we designate by $r$ the radius of the circumference described by any point, and by $c$ the corresponding value of $z$, we shall have

$$z = c, \quad (3) \qquad x^2 + y^2 = r^2. \quad (4)$$

But since the point is on the generatrix, every value $z = c$, must satisfy, at the same time, equations (1) and (2). If, therefore, the four equations be combined, the resulting equation will be the equation of the surface.

40. If it be required to find the surface described by a straight line revolving around the axis of $Z$, the equations of the generatrix will be

$$x = az + \alpha, \qquad y = bz + \beta,$$

and the equations of the circumference described by any

point of the line will be

$$z = c, \qquad x^2 + y^2 = r^2.$$

But every value of $c$ must satisfy the two first equations; hence,

$$x = ac + \alpha, \qquad y = bc + \beta.$$

and therefore,

$$(ac + \alpha)^2 + (bc + \beta)^2 = r^2.$$

Substituting for $c$ and $r$ their general values, and we obtain for the equation of the surface

$$(az + \alpha)^2 + (bz + \beta)^2 = x^2 + y^2.$$

41. If we assume, as we are at liberty to do, the axis of $X$ perpendicular both to the axis of revolution and the generatrix, the generatrix will then be parallel to the coördinate plane of $YZ$, and its equation will become

$$x = \alpha, \qquad y = bz,$$

and we shall also have

$$a = 0, \qquad \beta = 0.$$

Substituting these values of $a$ and $\beta$ in the equation of the surface, before found, and it becomes

$$x^2 + y^2 - b^2z^2 = \alpha^2,$$

which is the equation of a hyperboloid of revolution of one nappe (Art. 37).

## *Discussion of the General Equation of the Second Degree between three Variables.*

42. The general equation of the second degree, will be of the form

$$Az^2 + By^2 + Cx^2 + Dzy + Ezx + Fyx + A'z + B'y + C'x + L = 0. \quad (1)$$

If now, we change the direction of the co-ordinate axes without removing the origin, we shall have the formulas (Bk VIII, Prop. XVII.

$$x = x' \cos X + y' \cos X' + z' \cos X'',$$
$$y = x' \cos Y + y' \cos Y' + z' \cos Y'',$$
$$z = x' \cos Z + y' \cos Z' + z' \cos Z''.$$

If these values of the variables be substituted in equation (1) it will take the form

$$A'z'^2 + B'y'^2 + C'x'^2 + D'z'y' + E'z'x' + F'y'x' + A''z' + B''y' + C''x' + L' = 0,$$

in which the coefficients are functions of the angles

$$X,\ X',\ X''; \qquad Y,\ Y'\ Y''; \qquad Z,\ Z',\ Z''.$$

If it be required that the new co-ordinate axes shall be at right angles to each other, we shall have (Bk. VIII, Prop XVII, Sch. 5),

$$\cos X \cos X' + \cos Y \cos Y' + \cos Z \cos Z' = 0,$$
$$\cos X \cos X'' + \cos Y \cos Y'' + \cos Z \cos Z'' = 0,$$
$$\cos X' \cos X'' + \cos Y' \cos Y'' + \cos Z' \cos Z'' = 0$$

But we also have,

$$\cos^2 X + \cos^2 Y + \cos^2 Z = 1,$$
$$\cos^2 X' + \cos^2 Y' + \cos^2 Z' = 1,$$
$$\cos^2 X'' + \cos^2 Y'' + \cos^2 Z'' = 1.$$

The last six equations, are equations of condition between the nine angles which the new axes form with the primitive. But since there are nine arbitrary constants, three additional equations of condition may be established. We may, therefore, make in the transformed equation

$$D' = 0, \qquad E' = 0, \qquad F' = 0.$$

If it be required to find the values of these coefficients in terms of the angles, it may be observed that the terms of the general equation which contain the first powers of the variables, will contain no part of the coefficients $D'$, $E'$, $F'$ Without effecting the entire operations, we may readily form the coefficients, which being placed equal to 0, give the following equations,

$$\left.\begin{array}{l} 2A\cos Z\cos Z' + D(\cos Z\cos Y' + \cos Y\cos Z') \\ +2B\cos Y\cos Y' + E(\cos Z\cos X' + \cos X\cos Z') \\ +2C\cos X\cos X' + F(\cos Y\cos X' + \cos X\cos Y') \end{array}\right\} = 0,$$

$$\left.\begin{array}{l} 2A\cos Z\cos Z'' + D(\cos Z\cos Y'' + \cos Y\cos Z'') \\ +2B\cos Y\cos Y'' + E(\cos Z\cos X'' + \cos X\cos Z'') \\ +2C\cos X\cos X'' + F(\cos Y\cos X'' + \cos X\cos Y'') \end{array}\right\} = 0,$$

$$\left.\begin{array}{l} 2A\cos Z'\cos Z'' + D(\cos Z'\cos Y'' + \cos Y'\cos Z'') \\ +2B\cos Y'\cos Y'' + E(\cos Z'\cos X'' + \cos X'\cos Z'') \\ +2C\cos X'\cos X'' + F(\cos Y'\cos X'' + \cos X'\cos Y'') \end{array}\right\} = 0.$$

These three equations together with the six before established, will enable us to determine the nine angles which the new co-ordinate axes form with the primitive.

Introducing the conditions

$$D' = 0, \qquad E' = 0, \qquad F' = 0,$$

into the transformed equation, and omitting the accents, it takes the form

$$Az^2 + By^2 + Cx^2 + A'z + B'y + C'x + L = 0; \quad (2)$$

and since the above transformations are always possible, the last equation is as general as equation (1) from which it was derived.

43. Let us now transfer the origin of co-ordinates without changing the direction of the axes. For this, we have the formulas

$$x = a + x', \qquad y = b + y', \qquad z = c + z'.$$

Substituting these values of the variables in the last equation, and placing the known terms equal to $P$, that is, making

$$P = Ac^2 + Bb^2 + Ca^2 + A'c + B'b + C'a + L,$$

and omitting the accents, we obtain

$$Az^2 + By^2 + Cx^2 + \left.\begin{array}{c}2Ac \\ + A'\end{array}\right| z + \left.\begin{array}{c}2Bb \\ + B'\end{array}\right| y + \left.\begin{array}{c}2Ca \\ + C'\end{array}\right| x + P = 0$$

Since there are three arbitrary constants, $a$, $b$, $c$, introduced into the equation, such values may be attributed to them as to render

$$2Ac + A' = 0, \qquad 2Bb + B' = 0, \qquad 2Ca + C' = 0,$$

from which we obtain

$$c = -\frac{A'}{2A}, \qquad b = -\frac{B'}{2B}, \qquad a = -\frac{C'}{2C}.$$

These values will be real and finite, and consequently the transformation possible, when neither $A$, $B$, nor $C$ is 0: that is, when the second powers of the three variables enter into the transformed equation (2). The equation will then take the form

$$Az^2 + By^2 + Cx^2 + P = 0.$$

44. Let us now suppose that the coefficient of the second power of one of the variables, that of $x$ for example, should become equal to 0. We shall then have $C = 0$, and equation (2) will take the form

$$Az^2 + By^2 + A'z + B'y + C'x + L = 0. \quad (3)$$

If we again transfer the origin of co-ordinates, without changing the direction of the axes, we shall have the formulas

$$x = a + x, \qquad y = b + y, \qquad z = c + z,$$

in which we have omitted the accents.

These values being substituted in equation (3), and the coefficients of the first powers of $z$ and $y$ placed equal to 0, gives

$$2Ac + A' = 0 \quad \text{and} \quad 2Bb + B' = 0.$$

But these two equations introduce but two conditions, and since there are three arbitrary constants, another condition may yet be introduced. Let us therefore make the known terms, which we before represented by $P$, equal to 0. This will give

$$Ac^2 + Bb^2 + A'c + B'b + C'a + L = 0.$$

Hence, we have,

$$c = -\frac{A'}{2A}, \quad b = -\frac{B'}{2B}, \quad a = -\frac{(Ac^2 + Bb^2 + A'c + B'b + L)}{C'},$$

and these values will be real and finite, unless $C' = 0$, and the equation will take the form

$$Az^2 + By^2 + C'x = 0.$$

**45.** If we have, at the same time, $C = 0$, and $C' = 0$, the last transformation would be impossible, and equation (2) would take the form

$$Az^2 + By^2 + A'z + B'y + L = 0.$$

Since this equation contains but two variables, and *is true for all values of* x, it is the equation of a cylindrical surface, whose rectilinear elements are parallel to the axis of $X$, or perpendicular to the co-ordinate plane $YZ$ (Art. 15.). The base of the cylinder is an ellipse when the coefficients $A$ and $B$ have like signs, and an hyperbola when they have contrary signs.

**46.** If the coefficients of two of the terms involving the second powers of the variables should become equal to 0;

that is, if we have at the same time

$$B=0, \qquad C=0,$$

equation (2) would reduce to

$$Az^2+A'z+B'y+C'x+L=0.$$

It will not be necessary to transfer the origin of co-ordinates in order to determine the form of the surface represented by this equation. For, if we intersect the surface by a series of planes parallel to the co-ordinate plane $YX$, their equations will be

$$z=c, \qquad z=c', \qquad z=c'', \quad \&c. \quad \&c.;$$

and if these equations be combined with that of the surface the resulting equations will represent the lines of intersection Combining the equations, we find

$$B'y+C'x=L', \quad B'y+C'x=L'', \quad B'y+C'x=L''', \ \&c.,$$

which represent the projections of the lines of intersection on the co-ordinate plane $YX$.

But these are the equations of parallel straight lines; hence, the surface is cylindrical, and the rectilinear elements parallel to the co-ordinate plane $YX$.

To find the trace of the surface on the co-ordinate plane $ZX$, make $y=0$, and we have for the equation of intersection

$$Az^2+A'z+C'x+L=0,$$

which is the equation of a parabola.

The intersection with the co-ordinate plane $ZY$, may be found by making $x=0$. The curve of intersection is a parabola whose equation is

$$Az^2+A'z+B'y+L=0.$$

The surface, therefore, represented by the equation

$$Az^2 + A'z + B'y + C'x + L = 0,$$

is a cylindrical surface with a parabolic base.

47. Since the general equation of the second degree between three variables is symmetrical with respect to the three co-ordinate axes, all the results which are deduced by suppositions made on the coefficients of either variable, would be equally true if made on the coefficients of either of the others.

48. We therefore see, that the equation of every surface of the second degree will belong to one or the other of the three following classes:

*First Class.*

$$Mz^2 + Ny^2 + Lx^2 + P = 0$$

*Second Class.*

$$Mz^2 + Ny^2 + L'x = 0.$$

*Third Class.*

$$Mz^2 + Ny^2 + M'z + N'y + D = 0,$$

$$Mz^2 + M'z + N'y + L'x + D = 0.$$

The equations of the third class have already been discussed. They have been found to represent the surfaces of cylinders having for bases an ellipse, an hyperbola, or a parabola. We have then only to consider the equations of the first and second classes.

The first of these equations containing only the squares of the variables and an absolute term, the form of the equation will not be altered by changing $+x$, $+y$, $+z$, into $-x$, $-y$, $-z$; and hence, every straight line drawn through the origin of co-ordinates, and terminated by the surface, will be bisected at the origin. The origin of co-ordinates is, therefore, called the *centre of the surface*.

49. A plane which bisects a system of chords drawn parallel to each other, and terminated by the surface, is called a *central plane*, and the surface is said to be symmetrical with respect to this plane.

If the equation of the first class of surfaces be resolved with respect to each of the variables in succession, we shall find for every value of either, a corresponding equal value with a contrary sign: hence, each of the co-ordinate planes will bisect all chords drawn parallel to its axis and terminated by the surface. Hence also, each of the co-ordinate planes is a central plane: the intersection of either two is a diameter of the surface, and the point at which the three intersect, is the centre.

50. The surfaces which are comprised in the second class have no centres, since they are symmetrical only with respect to two of the co-ordinate planes, $ZX$, $YX$. The first and second classes of surfaces are distinguished from each other by this striking characteristic, *every surface of the first class has a centre, while not one of the second class enjoys this property.*

## *Discussion of the First Class of Surfaces represented by the Equation.*

$$Mz^2 + Ny^2 + Lx^2 + P = 0.$$

51. It is plain, that every variation of the signs which can be made among the coefficients of the variables, will be embraced in one of the three following cases:

1st. *When the coefficients of the variables are all plus and* P *plus or minus.*

2d. *When two of the coefficients are plus, one minus, and* P *plus.*

3d. *When two of the coefficients are plus, one minus, and* P *minus.*

For, if all the coefficients are negative, change the sign of every term of the equation. We shall thus render all the coefficients positive, and the sign of $P$ will be plus or minus. Again, if two of the coefficients are negative and one positive, we may, in like manner, change the signs of all the terms, which will render two positive and one negative, and the sign of $P$ will still be plus or minus. These three cases will give rise to three species of surfaces of the first.

*First species, when* M, N, *and* L *are positive.*

52. Let us first suppose $P$ negative. The equation will then take the form

$$Mz^2 + Ny^2 + Lx^2 = P.$$

Let the surface be now intersected by three planes, respectively parallel to the co-ordinate planes. The equations of the secant planes will be

$$x = a,$$

$$y = b,$$

$$z = c.$$

Combining these equations with that of the surface, there results,

| | | |
|---|---|---|
| for | $x = a$ | $Mz^2 + Ny^2 = P - La^2,$ |
| " | $y = b$ | $Mz^2 + Lx^2 = P - Nb^2,$ |
| " | $z = c$ | $Ny^2 + Lx^2 = P - Mc^2;$ |

from which we see that the curves of intersection by planes parallel to the co-ordinate planes, are ellipses, which become imaginary when

$$a^2 > \frac{P}{L}, \qquad b^2 > \frac{P}{N}, \qquad c^2 > \frac{P}{M}:$$

that is, when $a$, $b$, $c$, are either positive or negative, and numerically greater than

$$\sqrt{\frac{P}{L}}, \qquad \sqrt{\frac{P}{N}}, \qquad \sqrt{\frac{P}{M}}.$$

The ellipses will reduce to a point when

$$a = \pm\sqrt{\frac{P}{L}}, \quad b = \pm\sqrt{\frac{P}{N}}, \quad c = \pm\sqrt{\frac{P}{M}};$$

since the intersections will then reduce to

$$Mz^2 + Ny^2 = 0, \qquad Mz^2 + Lx^2 = 0, \qquad Ny^2 + Lx^2 = 0,$$

which can only be satisfied for the values

$$x = 0, \qquad y = 0, \qquad z = 0.$$

The surface, therefore, which we are now considering, is limited in every direction, since it is inscribed in the parallelopipedon, of which the equations of the faces are

$$x = \pm\sqrt{\frac{P}{L}}, \qquad y = \pm\sqrt{\frac{P}{N}}, \qquad z = \pm\sqrt{\frac{P}{M}}.$$

This surface is called an ellipsoid.

53. To determine the curves in which the surface is intersected by the co-ordinate planes, and which are called the *principal sections of the surface*, it is only necessary to combine the equations of the planes with the equation of the surface.

| | | |
|---|---|---|
| For | $x = 0$, | $Mz^2 + Ny^2 = P$, |
| " | $y = 0$, | $Mz^2 + Lx^2 = P$, |
| " | $z = 0$, | $Ny^2 + Lx^2 = P$. |

And for the points at which the surface cuts the three axes, we obtain

$$y=0, \qquad z=0, \qquad Lx^2=P, \quad \text{whence} \quad x=\sqrt{\frac{P}{L}},$$

$$x=0, \qquad z=0, \qquad Ny^2=P, \quad \text{“} \quad y=\sqrt{\frac{P}{N}},$$

$$x=0, \qquad y=0, \qquad Mz^2=P, \quad \text{“} \quad z=\sqrt{\frac{P}{M}}.$$

The lines

$$B'B=2\sqrt{\frac{P}{L}}, \qquad EE'=2\sqrt{\frac{P}{N}}, \qquad DD'=2\sqrt{\frac{P}{M}},$$

are called the *principal axes of the ellipsoid.* If we introduce them into the equation of the surface, we shall have the equation of the ellipsoid referred to its centre and axes. Let us make

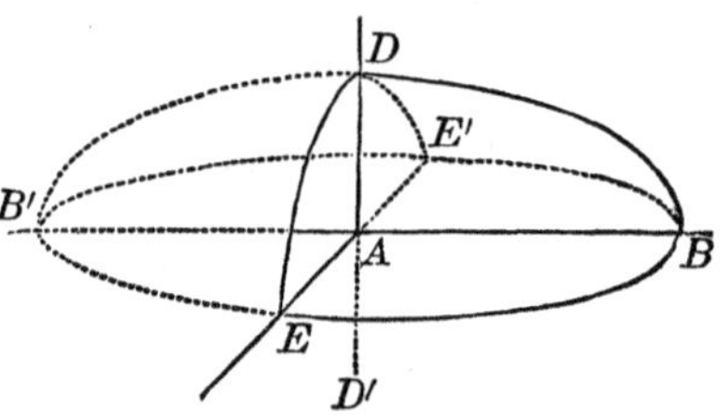


$$2A=2\sqrt{\frac{P}{L}}, \qquad 2B=2\sqrt{\frac{P}{N}}, \qquad 2C=2\sqrt{\frac{P}{M}};$$

from which we have

$$L=\frac{P}{A^2}, \qquad N=\frac{P}{B^2}, \qquad M=\frac{P}{C^2}.$$

Substituting these values in the equation of the surface and we obtain

$$A^2B^2z^2+A^2C^2y^2+B^2C^2x^2=A^2B^2C^2.$$

*Particular cases.*

**54.** Let us now suppose that the coefficients of either two of the variables become equal to each other, for example,

$N = L$: this will give $A = B$, and the equation of the surface will become

$$A^4z^2 + A^2C^2y^2 + A^2C^2x^2 = A^4C^2;$$

or dividing by $A^2$, we have

$$A^2z^2 + C^2y^2 + C^2x^2 = A^2C^2,$$

an equation which may be put under the form

$$y^2 + x^2 = \frac{A^2}{C^2}(C^2 - z^2), \quad \text{or} \quad y^2 + x^2 = F(z),$$

which is the equations of a surface of revolution, the axis coinciding with the axis of $Z$ (Art. 30).

If we make $z = \text{const} = c$, we shall have

$$x^2 + y^2 = \text{a constant};$$

which proves that every section, made by a plane perpendicular to the axis of $Z$, is a circle.

If we make, in succession, $y = 0$, and $x = 0$, in the equation of the surface, we shall obtain

$$A^2z^2 + C^2x^2 = A^2C^2, \qquad A^2z^2 + B^2y^2 = A^2B^2,$$

which are the equations of the intersections of the surface by the co-ordinate planes $ZX$, and $ZY$.

If we make $M = N$, or $M = L$, the surface will become a surface of revolution, the axis coinciding in the first case with the axis of $X$, and in the second with the axis of $Y$.

55. If we suppose $M = N = L$, we shall have $A = B = C$, and the equation of the surface will reduce to

$$z^2 + y^2 + x^2 = A^2,$$

which is the equation of a spherical surface, of which the centre is at the origin of co-ordinates (Art. 7).

56. Thus far, in the discussion, we have supposed the absolute term $P$ to be finite and negative. Let us now suppose, in the first place, that $P=0$; and secondly, that $P$ is positive.

The first supposition reduces the general equation to

$$Mz^2 + Ny^2 + Lx^2 = 0,$$

which can only be satisfied by the values

$$z=0, \qquad y=0, \qquad x=0;$$

and hence, this supposition reduces the surface to a *point.*

Under the second supposition, the equation of the surface takes the form

$$Mz^2 + Ny^2 + Lx^2 + P = 0,$$

in which all the terms are positive. Hence, the equation cannot be satisfied for real values of the variables; and therefore the *surface becomes imaginary.*

Hence we conclude, that the ellipsoid, represented by the equation

$$Mz^2 + Ny^2 + Lx^2 + P = 0,$$

has four varieties, viz:

1st. The ellipsoid of revolution.

2d. The sphere.

3d. The point.

4th. The imaginary surface.

*Second species, in which* M *and* N *are positive,* L *negative, and* P *positive.*

57. These suppositions will reduce the general equation to the form

$$Mz^2 + Ny^2 - Lx^2 = -P.$$

If we now intersect the surface by planes, respectively parallel to the co-ordinate planes, we shall have

$$\text{for} \quad x = a, \quad Mz^2 + Ny^2 = La^2 - P, \quad (1)$$

$$\text{"} \quad y = b, \quad Mz^2 - Lx^2 = -(Nb^2 + P), \quad (2)$$

$$\text{"} \quad z = c, \quad Ny^2 - Lx^2 = -(Mc^2 + P). \quad (3)$$

Equations (2) and (3) indicate that all sections made in the surface by planes parallel to the co-ordinate planes $ZX$ and $YX$ are hyperbolas, having their transverse axes parallel to the axis of $X$.

Equation (1) represents an ellipse. The curve will be real when

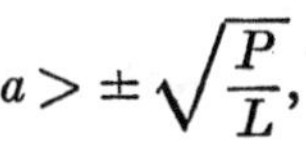

$$a > \pm\sqrt{\frac{P}{L}},$$

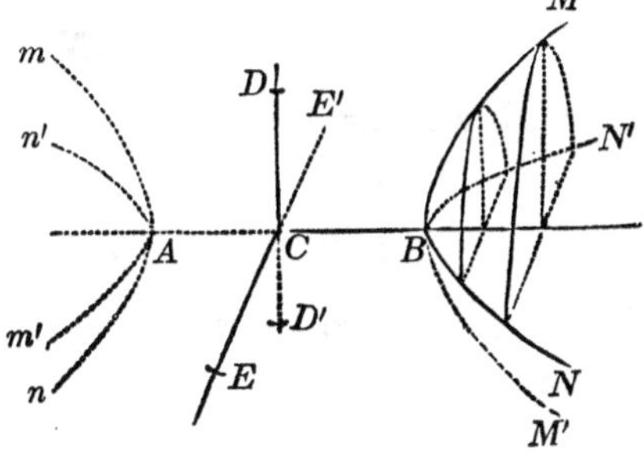


and imaginary when

$$a < \pm\sqrt{\frac{P}{L}}.$$

If, therefore, we make

$$CB = +\sqrt{\frac{P}{L}}, \quad \text{and} \quad CA = -\sqrt{\frac{P}{L}},$$

we see that the curve will be imaginary for all planes passing between the points $A$ and $B$, and real for all planes which are passed at a greater distance from the origin, either on the positive or negative side of abscissas. The sections on different sides of the origin, and equally distant from it, are equal to each other We therefore conclude, that the surface is composed of two opposite and equal branches; and since the sections in two directions are hyperbolas, the surface is called, *an hyperboloid of two nappes.*

The principal sections are obtained by making, in succes

sion, $x=0$, $y=0$, $z=0$: they are

$$Mz^2+Ny^2=-P,$$

$$Mz^2-Lx^2=-P,$$

$$Ny^2-Lx^2=-P.$$

The first section is an imaginary curve; the two others are the hyperbolas $MBM'$, $mAm'$; $NBN'$, $nAn'$, of which $AB$ is a common transverse axis.

We find for the axes of the surface

$$2A=2\sqrt{\frac{P}{L}},\quad 2B\sqrt{-1}=2\sqrt{\frac{-P}{N}},\quad 2C\sqrt{-1}=2\sqrt{\frac{-P}{M}};$$

from which we obtain

$$L=\frac{P}{A^2},\qquad N=\frac{P}{B^2},\qquad M=\frac{P}{C^2};$$

Substituting these values in the equation of the surface, it becomes

$$A^2B^2z^2+A^2C^2y^2-B^2C^2x^2=-A^2B^2C^2.$$

The line $AB$, which is represented by $2A$, is called the *transverse axis of the surface;* and $EE'$, $DD'$, which are represented by $2B\sqrt{-1}$, and $2C\sqrt{-1}$, are called *conjugate axes.* The transverse axis is real, and the conjugate axes are both imaginary.

*Third species, in which* M *and* N *are positive,* L *negative, and* P *negative.*

58. Under these suppositions, the equation of the surface will take the form

$$Mz^2+Ny^2-Lx^2=+P.$$

Intersecting the surface by planes, respectively parallel to the co-ordinate planes, we obtain,

$$\text{for} \quad x = a, \qquad Mz^2 + Ny^2 = La^2 + P, \qquad (1)$$
$$\text{"} \quad y = b, \qquad Mz^2 - Lx^2 = -Nb^2 + P, \qquad (2)$$
$$\text{"} \quad z = c, \qquad Ny^2 - Lx^2 = -Mc^2 + P. \qquad (3)$$

The first equation represents an ellipse which is always real, whatever be the value of $a$. The two others represent hyperbolas, in which the transverse axes will be parallel to the axes of $X$ when

$$b^2 > \frac{P}{N}, \qquad c^2 > \frac{P}{M}.$$

The transverse axis of the curve represented by equation (2), will be parallel to the axis of $Z$ when $b^2 < \frac{P}{N}$, and the transverse axis in equation (3), will be parallel to the axis of $Y$ when $c^2 < \frac{P}{M}$.

59. We shall obtain the principal sections of the surface, by making $a = 0$, $b = 0$, $c = 0$, which gives,

$$Mz^2 + Ny^2 = P, \qquad Mz^2 - Lx^2 = P, \qquad Ny^2 - Lx^2 = P.$$

The ellipse $BFC$, represented by the first equation, is the smallest ellipse which can be obtained by intersecting the surface by a plane parallel to the plane $YZ$.

The two other sections are hyperbolas. The transverse axis, $C'C$, of the one coincides with the axis of $Z$; and $BB'$, the transverse axis of the other, coincides with the axis of $Y$. The conjugate axes coincide with the axis of $X$, and are both imaginary.

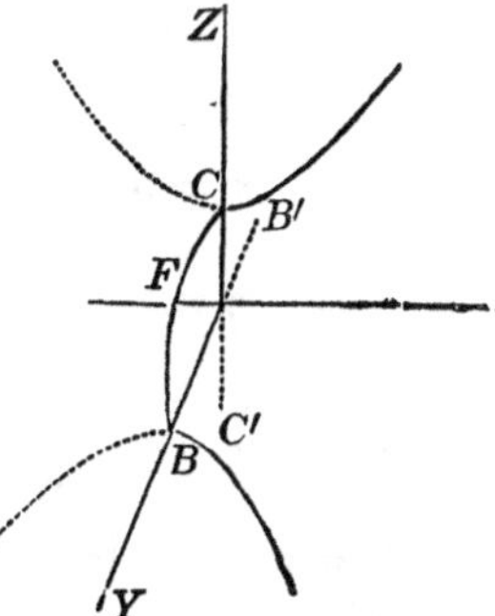


We find for the axes of the hyperbolas

$$2A\sqrt{-1}=2\sqrt{\frac{-P}{L}},\quad 2B=2\sqrt{\frac{P}{N}},\quad 2C=2\sqrt{\frac{P}{M}};$$

whence, $$L=\frac{P}{A^2},\quad N=\frac{P}{B^2},\quad M=\frac{P}{C^2}.$$

Substituting these values, and the equation of the surface reduces to

$$A^2B^2z^2+A^2C^2y^2-B^2C^2x^2=A^2B^2C^2.$$

The surface represented by this equation, is called the *hyperboloid of one nappe.*

*Particular cases of the Hyperboloids.*

60. If in the general equation we make $M=N$, which gives $C=B$, the equations of the surfaces become

$$A^2z^2+A^2y^2-B^2x^2=-A^2B^2,$$
$$A^2z^2+A^2y^2-B^2x^2=+A^2B^2;$$

which gives $$z^2+y^2=F(x):$$

hence, the two surfaces become surfaces of revolution, of which $X$ is the axis.

61. If we make $P=0$, the general equation will become

$$Mz^2+Ny^2-Lx^2=0.$$

The intersections of this surface by the co-ordinate planes will be,

for $x=0$, we have $y=0$, $z=0$,

" $y=0$, $z=\pm x\sqrt{\frac{L}{M}}$,

" $z=0$, $y=\pm x\sqrt{\frac{L}{N}}$;

hence, the surface is conical, the vertex being at the origin of co-ordinates

Therefore, the surfaces represented by the equation

$$Mz^2 + Ny^2 - Lx^2 \pm P = 0$$

are,

1st The hyperboloid of two nappes.

2d. The hyperboloid of one nappe.

These surfaces have two varieties, viz: the hyperboloids of revolution, and the conical surface.

## *Second Class of Surfaces in which*

$$Mz^2 + Ny^2 + L'x = 0.$$

62. The surfaces of this class are divided into two species,

1st. When $M$ and $N$ have like signs.

2d. When $M$ and $N$ have unlike signs.

If $M$ and $N$ have like signs, they may both be rendered positive, in which case $L'$ will be plus or minus.

### *First species, in which* M *and* N *are both positive.*

63. Let us suppose, in the first place, that $L'$ is negative. The equation will then take the form

$$Mz^2 + Ny^2 = L'x$$

If we intersect the surface by planes parallel to the co-ordinate planes, we shall have

for $x = a$, $Mz^2 + Ny^2 = L'a$,

" $y = b$, $Mz^2 = L'x - Nb^2$,

" $z = c$, $Ny^2 = L'x - Mc^2$.

22*

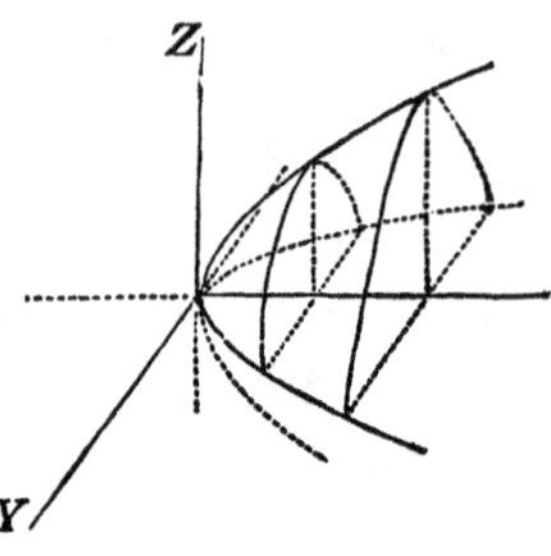


The first is the equation of an ellipse, which is always real when $a$ is positive, which reduces to a point when $a=0$, and becomes imaginary when $a$ is negative. Hence, the surface extends indefinitely in the direction of the positive abscissas, and is limited in the opposite direction by the co-ordinate plane $YZ$.

The two other sections are parabolas whose principal axes are parallel to the axis of $X$, and extend indefinitely in the direction of the positive abscissas.

If we make $b=0$, $c=0$, the equations of the parabolas become

$$z^2 = \frac{L'}{M}x, \qquad y^2 = \frac{L'}{N}x.$$

This surface is called an *elliptical paraboloid.*

64. If we suppose $L'$ to be positive, the equation of the surface will be of the form

$$Mz^2 + Ny^2 = -L'x.$$

If, in this equation, we change $x$ into $-x$, it will become

$$Mz^2 + Ny^2 = L'x,$$

the same equation as before discussed, excepting that $x$ is now negative where it was before positive.

The surface represented by the last equation will, therefore extend indefinitely in the direction of $x$ negative, and be limited in the other direction by the co-ordinate plane $YZ$

65. If we suppose $M=N$, the equation of the surface will become

$$z^2 + y^2 = \frac{L'}{M}x = F(x),$$

which is the equation of a paraboloid of revolution, of which $X$ is the axis (Art. 38).

*Second species, in which* M *is positive and* N *negative.*

66. Under this supposition, the equation becomes

$$Mz^2 - Ny^2 + L'x = 0.$$

It will only be necessary to consider the case in which $L'$ is negative; for if $L'$ were positive, the term may always be made negative by changing $x$ into $-x$, which merely changes the position of the surface with respect to the origin of co-ordinates, without affecting its properties.

If we suppose $L'$ negative, the equation becomes

$$Mz^2 - Ny^2 = L'x.$$

If the surface be now intersected by planes parallel to the co-ordinate planes, we shall have,

| | | |
|---|---|---|
| for | $x = a,$ | $Mz^2 - Ny^2 = L'a,$ |
| " | $y = b,$ | $Mz^2 = L'x + Nb^2,$ |
| " | $z = c,$ | $Ny^2 = -L'x + Mc^2.$ |

The two last equations represent parabolas, of which the principal axes are parallel to the axis of $X$. The axis of the first extends in the direction of the positive abscissas, and that of the second in the direction of the negative abscissas The parameters are $\dfrac{L'}{M}$ and $\dfrac{L'}{N}$.

The first equation represents an hyperbola of which the transverse axis is parallel to the axis of $Z$ when $a$ is positive, and parallel to the axis of $Y$ when $a$ is negative.

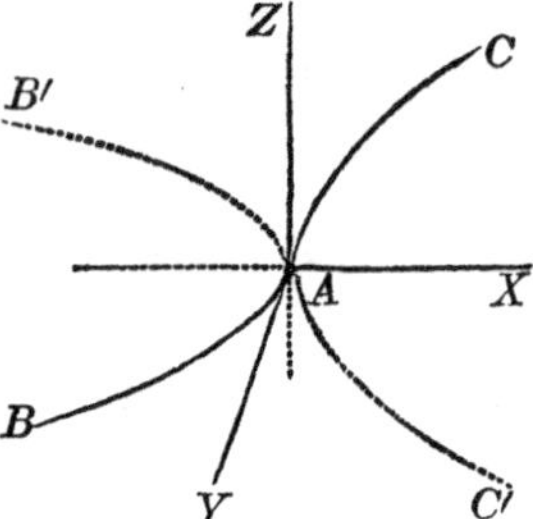


The two principal sections, by the co-ordinate planes $YX$, $ZX$, are the two parabolas, $BAB'$, $C'AC$, of which the equations are

$$Ny^2 = -L'x, \qquad Mz^2 = L'x.$$

The equation of the sections by the plane $YZ$, is

$$Mz^2 - Ny^2 = 0, \qquad \text{hence} \qquad z = \pm y\sqrt{\frac{N}{M}};$$

the sections, therefore, are two straight lines which intersect each other at the origin.

If we designate the ordinate of either line by $y'$. we have

$$Mz^2 - Ny'^2 = 0;$$

and subtracting this from the equation of intersection by a plane parallel to $YZ$, we shall have

$$Ny'^2 - Ny^2 = L'a,$$

or

$$y' - y = \frac{L'a}{N(y + y')};$$

hence, the straight lines whose equations are

$$z = \pm y\sqrt{\frac{N}{M}},$$

are asymptotes of the projections on the plane $YZ$ of all intersections by planes parallel to the co-ordinate plane $YZ$. Hence, if two planes be drawn through the axis of $X$, and these lines respectively, they will contain the surface, and become tangent to it at an infinite distance. They will therefore have the same relation to the surface which rectilineal asymptotes bear to the curves, to which they continually approach.

The sections of this surface being hyperbolas and parabolas, it is called an *hyperbolic-paraboloid.*

67. We conclude from the preceding discussion, that the general equation of the second degree represents five kinds of surfaces, viz:

1. *The Ellipsoid,* having for varieties, the ellipsoid of revolution, the sphere, a point, and the imaginary surface.

2. *The Hyperboloid of two Nappes,* which has for a variety the hyperboloid of revolution of two nappes; and the *Hyperboloid of one Nappe,* which has two varieties, the hyperboloid of revolution of one nappe, and the conic surface.

3. *The Elliptical Paraboloids,* having for a variety the paraboloid of revolution.

4. *The Hyperbolic Paraboloid,* which has no variety.

5. *The Cylindrical Surfaces,* with elliptical, hyperbolic, and parabolic bases. These last, however, are a species of paraboloids, since they are obtained by supposing that the second power of *one* or two of the variables, disappears in the first transformation of the general equation.

68. It may be remarked here, that the equation of a plane being of the first degree, will, when combined with the equation of a surface of the second order, give an equation of the second degree: hence, the intersection of any plane with a surface of the second order, will be a curve of the second order.

## *Of Tangent Planes to Surfaces of the Second Order.*

69. A plane is said to be tangent to a surface when there is at least one point common to the plane and surface, through which if any number of planes be drawn, the sections made in the plane will be tangent to the sections made in the surface. But two straight lines which intersect each other

determine the position of a plane. If, therefore, through any point of a surface we pass two planes, and draw a tangent line to each of the curves of intersection at their common point, the plane passing through these tangents will be tangent to the surface at the common point.

*Of tangent planes to surfaces which have a centre.*

70. The equation of surfaces having a centre is of the form

$$Mz^2 + Ny^2 + Lx^2 + P = 0.$$

Let $x''$, $y''$, $z''$, designate the co-ordinates of a point at which the plane is to be tangent to the surface. These co-ordinates will satisfy the equation of the surface, and give

$$Mz''^2 + Ny''^2 + Lx''^2 + P = 0.$$

Through the point of tangency, let two planes be drawn respectively parallel to $YZ$ and $ZX$. We shall then have,

for $\quad x = a'' = x'', \qquad Mz^2 + Ny^2 + La''^2 + P = 0,$

" $\quad y = b'' = y'', \qquad Mz^2 + Lx^2 + Nb''^2 + P = 0.$

But the equations of two straight lines, drawn tangent to these curves at their common point, of which the co-ordinates are $x''$, $y''$, $z''$, are

$$x = a'' = x'', \qquad Mzz'' + Nyy'' + La''^2 + P = 0,$$

$$y = b'' = y'', \qquad Mzz'' + Lxx'' + Nb''^2 + P = 0.$$

The equation $\qquad x = a'' = x'',$

represents the projection of the first tangent on the co-ordinate plane $ZX$; and the equation

$$y = b'' = y'',$$

represents the projection of the second tangent on the co-ordinate plane $YZ$. The projections of these tangents coincide, respectively, with the traces of the secant planes. The other two equations, being the common equations of tangent lines, are easily recognised.

The equations of these tangents may be placed under the forms,

$$x = 0 \, . \, z + a'', \qquad y = -\frac{Mz''}{Ny''} \, . \, z \quad -\frac{(La''^2 + P)}{Ny''}.$$

$$x = -\frac{Mz''}{Lx''} z - \frac{(Nb''^2 + P)}{Lx''}, \qquad y = 0 \, . \, z + b''.$$

comparing these with the general equations of two straight lines, we find, for the first,

$$a = 0, \qquad b = -\frac{Mz''}{Ny''};$$

and for the second,

$$a' = -\frac{Mz''}{Lx''}, \qquad b' = 0.$$

It is now required to find the equation of a plane which shall pass through these two tangent lines.

The equation of a plane passing through the point of tangency, is of the form

$$A(x - x'') + B(y - y'') + C(z - z'') = 0;$$

and this plane will contain both of the tangents (Bk. VIII, Prop. XII), if

$$Aa + Bb + C = 0,$$

and $$Aa' + Bb' + C = 0.$$

Substituting in these equations the values of $a$, $b$, $a'$, $b'$, before found, and we have

$$B \times -\frac{Mz''}{Ny''} + C = 0, \quad \text{whence,} \quad B = \frac{Ny''}{Mz''} \; C,$$

$$A \times -\frac{Mz''}{Lx''} + C = 0, \quad \text{whence,} \quad A = \frac{Lx''}{Mz''} \; C.$$

Substituting these values in the equation of the plane, and we obtain

$$Mz''(z - z'') + Ny''(y - y'') + Lx''(x - x'') = 0,$$

or by reducing, and recollecting that

$$Mz''^2 + Ny''^2 + Lx''^2 = -P,$$

we obtain, for the equation of the tangent plane,

$$Mzz'' + Nyy'' + Lxx'' + P = 0.$$

*Of tangent planes to surfaces which have not a centre.*

70. The equation of the paraboloids is

$$Mz^2 + Ny^2 + 2L''x = 0,$$

by making $2L'' = L'$.

If we designate the co-ordinates of the point of tangency by $x''$, $y''$, $z''$, we shall have.

$$Mz''^2 + Ny''^2 + 2L''x'' = 0.$$

The equations of the curves of intersection, by planes passing through the point of tangency and respectively parallel to the co-ordinate planes $YZ$, $ZX$, are

$$x = a'' = x'', \qquad Mz^2 + Ny^2 + 2L''a'' = 0,$$

$$y = b'' = y'', \qquad Mz^2 + 2L''x + Nb''^2 = 0.$$

The equations of the tangents to these curves, at their common point, are

$$x = a'' = x'' \qquad Mzz'' + Nyy'' + 2L''a'' = 0,$$

$$y = b'' = y'' \qquad Mzz'' + L''(x + x'') + Nb''^2 = 0.$$

Comparing these with the general equations of two lines, and we find

$$a=0, \qquad b=-\frac{Mz''}{Ny''}, \qquad a'=-\frac{Mz''}{L''}, \qquad b'=0.$$

The equation of a plane passing through the point of tangency, is of the form

$$A(x-x'')+B(y-y'')+C(z-z'')=0;$$

and this plane will contain both the tangents, if

$$Aa+Bb+C=0, \qquad \text{and} \qquad Aa'+Bb'+C=0.$$

Substituting the values before found for $a$, $b$, $a'$, $b'$, and we have,

$$B\times-\frac{Mz''}{Ny''}+C=0; \qquad \text{whence,} \qquad B=\frac{Ny''}{Mz''}.C,$$

$$A\times-\frac{Mz''}{L''}+C=0; \qquad \text{whence,} \qquad A=\frac{L''}{Mz''}.C.$$

If these values be substituted in the equation of the plane, it becomes

$$Mz''(z-z'')+Ny''(y-y'')+L''(x-x'')=0;$$

or by reducing, and recollecting that the co-ordinates of the point of tangency satisfy the equation of the surface, it becomes

$$Mzz''+Nyy''+L''(x+x'')=0.$$

72. If a line be drawn perpendicular to a tangent plane at the point of contact, it is said to be normal to the surface, and is called a *normal line*.

If the equation of a normal be required, and the tangent plane and the point of contact are known, the problem consists in finding the equations of a line which shall be perpendicular to a given plane at a given point.

The equations of a normal, in the first class of surfaces, are

$$x - x'' = \frac{Lx''}{Mz''}(z - z''), \qquad y - y'' = \frac{Ny''}{Mz''}(z - z'');$$

and in the second class,

$$x - x'' = \frac{L''}{Mz''}(z - z''), \qquad y - y'' = \frac{Ny''}{Mz''}(z - z'').$$

## *Of the Generation of the Hyperboloid of one Nappe and of the Hyperbolic Paraboloid, by the movement of a straight line.*

73. It is now proposed to show, that the hyperboloid of one nappe and the hyperbolic paraboloid, may each be generated in two different ways by the movement of a straight line.

74. Let us resume the equation of surfaces which have a centre,

$$Mz^2 + Ny^2 + Lx^2 + P = 0. \quad (1)$$

The equation of a tangent plane is

$$Mzz'' + Nyy'' + Lxx'' + P = 0; \quad (2)$$

and since the co-ordinates of the point of tangency satisfy the equation of the surface, we also have

$$Mz''^2 + Ny''^2 + Lx''^2 + P = 0. \quad (3)$$

In order to determine the points which are common to the tangent plane and surface, we must combine their equations.

Now, if twice equation (2) be subtracted from the sum of equations (1) and (3), the result may be placed under the form,

$$M(z - z'')^2 + N(y - y'')^2 + L(x - x'')^2 = 0, \quad (4)$$

in which $x$, $y$, $z$, are the co-ordinates of all the points which are common to the tangent plane and surface.

If the coefficients $M$, $N$, $L$, are positive, the surface repre sented by equation (1), will be an ellipsoid (Art. 53). Under this supposition, equation (4) can only be satisfied by making

$$x = x'', \qquad y = y'', \qquad \text{and} \qquad z = z'';$$

and hence, *the ellipsoid and its tangent plane have but a single point in common.*

75. Let us now suppose $M$ and $N$ positive, and $L$ negative. Under this supposition, equation (4), which represents the points common to the tangent plane and surface, will become

$$M(z - z'')^2 + N(y - y'')^2 - L(x - x'')^2 = 0. \quad (5)$$

The equation of the tangent plane

$$Mzz'' + Nyy'' - Lxx'' + P = 0,$$

may be placed under the form,

$$Mz''(z - z'') + Ny''(y - y'') - Lx''(x - x'') = 0. \quad (6)$$

It is now required to find all the values of $x$, $y$, and $z$, which will satisfy equations (5) and (6).

For this purpose, let us combine with each, the equations of a straight line passing through the point of tangency. These equations are,

$$x - x'' = a(z - z''), \qquad y - y'' = b(z - z''). \quad (7)$$

If now we combine equations (5), (6), and (7), and find single and real values for $a$ and b, the surface and tangent plane will have one straight line in common; if $a$ and $b$ have two values, they will have two lines in common; and if $a$ and $b$ are indeterminate, there will be an infinite number of straight lines common with the plane and surface

Substituting the values of $x-x''$, $y-y''$, in equations (5) and (6), they become

$$(z-z'')^2(M+Nb^2-La^2)=0,$$
$$(z-z'')\,(Mz''+Nby''-Lax'')=0;$$

or by dividing the equations respectively by $(z-z'')^2$ and $z-z''$,

$$M+Nb^2-La^2=0, \quad (8)$$
$$Mz''+Nby''-Lax''=0. \quad (9)$$

which are the equations of condition in order that the straight line shall coincide with the tangent plane and surface. Let us now see if these equations will give real values for $a$ and $b$.

We find from equation (9),

$$a=\frac{Ny''b+Mz''}{Lx''}. \quad (10)$$

Substituting this value in equation (8), and reducing

$$N(Lx''^2-Ny''^2)b^2-2MNy''z''b=M^2z''^2-MLx''^2,$$

whence, $b=\dfrac{MNy''z''\pm\sqrt{MNLx''^2(Mz''^2+Ny''^2-Lx''^2)}}{N(Lx''^2-Ny''^2)}$

But, $Mz''^2+Ny''^2-Lx''^2=P$, hence,

$$b=\frac{MNy''z''\pm x''\sqrt{MNL\,.\,P}}{N(Lx''^2-Ny''^2)}.$$

Since the negative sign belonging to $L$ has already been attributed to it, and since $M$ and $N$ are both positive, the quantity under the radical will be negative when $P$ is negative, and positive when $P$ is positive. But when $P$ is negative in the second member, the surface is the hyperboloid of two nappes (Art. 57); and when $P$ is positive, it is the hyperboloid of one nappe (Art. 58). In the latter case, $b$ *has two real values;* and if these values be substituted in equation (10), we shall find two real values for $a$. Hence *if a tangent plane be drawn to the hyperboloid of one nappe, at any point of the surface, it will contain two straight lines common to the plane and surface, which intersect each other at the point of contact.*

The surface of the hyperboloid may therefore be generated by the movement of either of these lines; hence we say, that it has *two generations.*

76. Let us now pass to the paraboloids, of which the equation is

$$Mz^2 + Ny^2 + 2L''x = 0; \quad (1)$$

and the equation of the tangent plane,

$$Mzz'' + Nyy'' + L''(x + x'') = 0. \quad (2)$$

But since the point of contact is on the surface, we have

$$Mz''^2 + Ny''^2 + 2L''x'' = 0. \quad (3)$$

If we add equations (1) and (3), and from the sum subtract twice equation (2), we shall find

$$M(z - z'')^2 + N(y - y'')^2 = 0. \quad (4)$$

Now, if $M$ and $N$ are both positive, which supposition gives the elliptical paraboloid (Art. 63), there will be but *one point* common to the tangent plane and surface.

But if we take the hyperbolic paraboloid, in which $N$ is negative, equation (4) will become

$$M(z - z'')^2 - N(y - y'')^2 = 0; \quad (5)$$

whence, $$y - y'' = \pm (z - z'')\sqrt{\frac{M}{N}},$$

from which we see, that equation (5) represents the projections on the co-ordinate plane $YZ$ of two straight lines which are common to the surface of the hyperbolic paraboloid and the tangent plane, and which intersect each other at the point of contact.

To determine the directions of these lines, we will place the equation of the tangent plane under the form

$$Mz''(z-z'')+Ny''(y-y'')+L''(x-x'')=0. \quad (6)$$

Let us now combine the equations

$$x-x''=a(z-z''), \qquad y-y''=b(z-z''), \quad (7)$$

of a straight line passing through the point of contact with equations (5) and (6). After dividing by the common factor $(z-z'')^2$, we find

$$M-Nb^2=0, \qquad Mz''-Nby''+L''a=0;$$

whence,

$$b=\pm\sqrt{\frac{M}{N}},$$

and

$$a=\frac{Mz''\mp y''\sqrt{MN}}{L''},$$

values of $a$ and $b$, which are always *real*, when the surface is the hyperbolic paraboloid. Hence, *two straight lines may always be drawn through any point on the surface of a hyperbolic paraboloid, which shall coincide with the surface, and therefore the surface may be generated by the movement of either one of two right lines.*

If we substitute, in equation (7), the value of $b$, we find

$$y-y''=\pm(z-z'')\sqrt{\frac{M}{N}},$$

which corresponds with equation (5).

Since $b$ is independent of the co-ordinates of the point of contact, $x''$, $y''$, $z''$, it follows that the projections of all the elements of each generation on the co-ordinate plane $YZ$ are parallel to each other: hence, the elements of the same generation are situated in a series of parallel planes, and are consequently parallel to a given plane, which is called the *plane-director.*

www.ingramcontent.com/pod-product-compliance
Lightning Source LLC
LaVergne TN
LVHW010158110826
845151LV00002B/549

* 9 7 8 1 4 2 5 5 3 7 3 8 8 *